THE NUCLEAR APPLE

By the same Author

*

INTRODUCTION
TO QUANTUM MECHANICS
(*McGraw-Hill*)

THE NUCLEAR APPLE

Recent Discoveries in Fundamental Physics

P. T. MATTHEWS, F.R.S.
Professor of Theoretical Physics
Imperial College of Science and Technology
University of London

The universe is not to be narrowed down to the limits of the understanding, as has been man's practice up till now, but rather the understanding must be stretched and enlarged, to take in the image of the universe as it is discovered.
FRANCIS BACON

ST. MARTIN'S PRESS
NEW YORK

For information, write:
St. Martin's Press, Inc., 175 Fifth Ave., New York, N.Y. 10010
Printed in Great Britain
Library of Congress Catalog Card Number: 75-173566
First published in the United States of America in 1971

AFFILIATED PUBLISHERS: Macmillan & Company, Limited, London—also at Bombay, Calcutta, Madras and Melbourne—The Macmillan Company of Canada, Limited, Toronto

Contents

Illustrations

Acknowledgements

A number of friends and colleagues have helped me with this book. Particularly, I should like to thank Victor Weisskopf for permission to quote from the Pauli letter, and Peter Astbury, David Binnie, Stephen Goldsack, Owen Lock and Douglas Merrison for assistance with the photographs. I should also like to thank David Bartlett for some of the drawings and Susan Bartlett, Mavis Moriarty and Sally Staniland for their work on the manuscript in various stages of its development.

Introduction

Very dramatic discoveries were made in fundamental physics during the first thirty years of this century. These included the formulation of the theory of relativity, which is primarily concerned with the curious ramifications that follow from the finiteness and universality of the velocity of light. The period also covers the development of quantum mechanics, which explained the behaviour of atoms and their interaction with radiation. The general ideas of these developments were made available to non-specialists in a number of books, among which those of Sir Arthur Eddington and Sir James Jeans were outstanding.

Since the end of the Second World War there has been further development along this frontier of our understanding of the Universe which has now moved to nuclear and sub-nuclear phenomena. Here, for the first time, with the aid of the big proton and electron accelerators it has been possible to study physical situations right outside the domain of common experience in which the strange consequences of both relativity and quantum mechanics are to be found operating simultaneously with full force. A great deal of progress has been made, which has deepened our knowledge of the mysterious Universe and the nature of the physical world. Apart from sporadic efforts in the popular press, very little has been done to bring the significance of these new discoveries within the reach of people with a general scientific training but no specialist knowledge of nuclear physics. Indeed, increased specialisation has tended to make this whole field the preserve of those actually working in it. This book is an attempt to remedy this situation, and to discuss the implications of the discoveries of the post-war period in a language which does not assume an advanced scientific background.

As we have said, all this work is carried out in the relativistic, quantum regime. This has not led to any further generalisation of the framework of physics, but it has given rise, through familiarity, to a more confident understanding of the ideas which seemed extremely bizarre when they were first proposed by Einstein, Heisenberg and Schrödinger. We, therefore, also include a general discussion of relativity and quantum mechanics, and the effect which they have had on the rigid Laplacian determinism, which seems to emerge from the concepts of 19th-century physics. General topics which come up are the significance of the flow of time, the changing concept of the basic building blocks of the Universe and the nature and limitation of physical predictions. We emphasise particularly the strong cross-coupling between nuclear and sub-nuclear research and the interpretation of astrophysical observations.

The first chapter traces briefly the origins of modern ideas in the heroic age of Greek science, and summarises the extremely optimistic situation which existed after the discovery of relativity and quantum mechanics. The next chapter shows how this scheme proved totally inadequate to deal with even the simplest properties of atomic nuclei. The four following chapters go in more detail into the discoveries which have been made in nuclear and sub-nuclear physics during the last thirty years. These include the distinction between Strong and Weak nuclear interactions, the finding of a spectrum of sub-nuclear particles, the three charges, the breakdown of space and time reversal invariance and the prediction of Ω^-. Evidently a whole new field has been opened up revealing a richness and diversity which has taken us completely by surprise and significantly altered our outlook on the physical world. This general question of outlook is developed in the last chapter, where we discuss the nature and limitations of physical predictions imposed by quantum and statistical mechanics. Finally, we briefly consider the relationship between life and thought and this general scheme.

1

Classical Physics, Relativity and the Quantum

The Origins of Modern Science

One of the salient features of everyday life is our inability to predict the future. No matter what precautions we take, tomorrow the storm may break or the child be run over in the street. This element of uncertainty is particularly strong in human affairs, as anyone knows who has tried to make a quick profit on the stock market, but the difficulty of predicting the weather shows that, from one hour to the next almost anything can happen even in a situation which is not subject in any way to human whim.

On the other hand it is clear that what takes place tomorrow is strongly linked to what is going on today, and the prospects of next year are definitely influenced by the events of this. Although it is difficult to make reliable forecasts of the weather, the seasons follow a regular pattern and the times of sunrise and sunset, and even eclipses, can be predicted with precision. To people accustomed to the vagaries of their local railway, one of the most impressive features of manned moon flights is the accuracy with which the planners can foretell the timing of each stage of the manoeuvre. Thus our consciousness of the uncertainty of the future is matched, and to some extent contradicted, by a compensating awareness of a time sequence of cause and effect, particularly in situations not involving mental processes. A kettle put on a fire will eventually boil; a tennis ball thrown against a wall will bounce back.

It is the function of the scientist, and particularly of the physicist, to identify those aspects of the physical world which are subject to precise prediction, and to give clear definitions of the properties of physical objects which play the crucial role in this predictable scheme. He tries to discover the laws, which relate these properties to each other and determine how they develop from one moment to the next.

The problem was formulated during the fifth and sixth centuries B.C. by Greek scientists, who also suggested a framework within which one could look for an answer. The earliest work was that of Thales of Miletus in Ionia, who can accurately be dated, because he predicted an eclipse of the Sun in 585 B.C. with the aid of Babylonian astronomical tables. Up to his time the forces at work in the World were thought of as generalisations of human emotions, and were explained by attributing to gods the powers of men and animals on a magnified scale. Thales was the first to develop a naturalistic cosmology in which the Universe was pictured as composed of some basic inanimate substance (water in his case), which moved under forces inherent in itself—not at the whim of the gods. This idea developed rapidly. Anaximander, also of Miletus and born before the death of Thales, replaced water as the basic material by a common indeterminate substance, from which he supposed the elements Earth, Mist, Fire and Water were formed. These 'make reparation and satisfaction to one another for their injustice according to the ordering of time'. It is evident that we have here notions which are more closely related to the different states of matter—solid, gas, liquid and radiation—than to the modern elements, and that the word 'justice' is used in a sense approximating to what we would now call 'equilibrium'. But the essential point is the continued development of a naturalistic approach.

In contrast to the Ionians, the school of Pythagoras, which existed at about the same time at Croton in Southern Italy, continued to follow the older religious tradition, giving preeminence to things of the soul, and believing that Nature was under the government of mind. The conflict of emphasis on mind or matter has echoed on down through the ages, and we shall meet it again. But the Pythagoreans were exceptional in that they had a deep mathematical sense, and in their interpretation of the World attached enormous importance to numbers, if not to measurement. This was expressed, for example, by Philolaus, a Pythagorean of the 5th century who wrote: 'Were it not for number and its nature, nothing that exists would be clear to anybody either in itself or in its relation to other things.' The Pythagoreans were fascinated by numerical patterns such as the series, ., .·., .·:·., which dominated their cosmology. Although most of modern science has been contrary to the Pythagorean tradition, the emphasis on mathematical patterns has reappeared in very recent discoveries in a remarkable way, which we describe in Chapter 6.

Important contributions to the naturalist way of thinking of the Ionian school were made by Empedocles of Sicily, who showed by experiment that air in general—not just mist—was a corporate substance, thus deducing the existence of something invisible from visible evidence, which is crucial to all modern science. But the real breakthrough came towards the end of the fifth century B.C. from the atomists Leucippus, another Ionian, and Democritus, a native of Abdera in Thrace, in an amazing anticipation of much of modern science. They proposed that everything is made of atoms of various shapes and sizes which have bulk, but are indestructable and infinite in number. The atoms move in an infinite void under the influence of natural laws.

The logic of this conception is that if you want to find how something works, in the causal sense defined above, the best thing to do is to take it to pieces. In modern terms this is fairly obvious with a motor-cycle or an alarm-clock, but the botanists and biologists have shown that the method also works for a flower or an earth-worm. The chemist takes the components to pieces to find molecules and atoms; the physicist then takes the atoms to pieces to find yet more basic elements. Each structure can be understood in terms of its component parts and the effects which they have on each other. As one works one's way downwards in size, the components at one level become the structures at the next. Ultimately one might hope to find constituent elements and mutual interactions which are so simple that they defy further analysis. In terms of these elementary constituents one should be able to describe, at least in principle, the causal aspect of the entire Universe, the whole system developing inexorably like some enormous machine.

The conceptual achievements of these early Greek thinkers are quite astounding and were far, far ahead of the necessary practical techniques for putting their ideas to experimental test. Most of the great men who followed them, such as Plato and Aristotle, were more impressed by Pythagorean concepts, with the emphasis on mind and the immortal soul. They made extensive advances in mathematics and observational astronomy, but their physical theories tended to be marred by a pseudo-mathematical religiosity. The naturalistic view of the Universe lay dormant for two thousand years. It was revived again during the Italian Renaissance, by such men as Leonardo da Vinci. There were major developments in the 17th century, and progress has been accelerating ever since, with the result that round about 1930 we seemed to be on the verge of a complete explanation of the physical world exactly along the lines which the atomists had proposed.

By that time it appeared to have been established that all matter in the Universe could be described in terms of just two types of indivisible particle—the *proton* and the *electron*—interacting with each other through two types of force. These are *electromagnetic* forces, whose effects were completely summarised during the 19th century by Maxwell in four fairly simple equations, and *gravitational* forces which operate according to the rules first appreciated two hundred years earlier by Newton, when he was struck by the famous apple. This is a fantastically simple synthesis of the vast variety of physical, chemical and biological phenomena. Let us review briefly how it works.

Electromagnetism, Gravity and the Quantum

The proton and electron each have mass, the proton being about two thousand times heavier than the electron. They also carry electric charges which are equal in magnitude, but opposite in sign. Maxwell's equations determine the forces which moving point charges exert on each other and show, in particular, that there will be an electrical attraction between the opposite charges on the proton and the electron. This makes possible the simplest atomic structure, the element of hydrogen. This consists of an electron whirling in an orbital about a central proton, much as a boy might whirl a stone round his head on the end of a string, the electrical attraction playing the role of the tension in the string. The next element, helium, has two electrons orbiting about a nucleus which contains two protons. The other chemical elements are built up simply by adding protons to the atomic nucleus and compensating electrons in the orbitals so that the total electric charge of the atom is always zero. Matter made of such atoms would be characterless and amorphous were it not for the effects of quantum mechanics. These were analysed in a great burst of activity between 1926 and 1928 by Schrödinger, Heisenberg and Dirac, following the earlier work of Planck and Bohr. We have a lot more to say about quantum mechanics in Chapter 7. At this stage we simply state the most important conclusions.

The main rotating shaft of an engine, for example a standard car engine, is usually attached to a flywheel. This has the property that it is hard to set in motion, but, once rotating, it tends to continue at the same speed, and thus smoothes out the jerky action which would otherwise arise from the successive firings in the cylinders. The property of the flywheel, which has this tendency to stay constant, depends on its mass, shape and the speed with which it is rotating, and is called the angular momentum. According to classical ideas the rate of rotation can be varied smoothly, so the angular momentum also varies continuously from one value to the next. But quantum mechanics has shown that this is not strictly true. Angular momentum can only be added or subtracted from a system in discrete jumps, which must be integer (or half integer) multiples of a fixed quantum, *h*, discovered by Planck.* Thus angular momentum varies not continuously, like the depth of a bath, but like a bank balance in which the least possible change is a penny added or subtracted. If we are dealing in hundreds of pounds the distinction between discrete and continuous change is negligible, but for a child whose pocket-money is threepence per week it makes all the difference in the world. This is exactly the physical situation. A shirt button rotating once per second—a pretty feeble flywheel—has an angular momentum in Planck's unit of some thousand million million million million (10^{27}) and adding or subtracting single quanta from this is not significantly different from a continuous change. However, the whirling electron in an atom may also be regarded as a miniature flywheel in which *h* is the natural unit to measure its angular momentum. Simple atoms have angular momenta of a few *h*, and one unit more or less is a very significant change.

Since the energy of the atom, like the energy stored in a spinning flywheel, depends on its mass, size and speed of rotation, the fact that the angular momentum is discrete implies that the energy can also only have a set of discrete values. The allowed values depend on a combination of the quantum conditions and Maxwell's electrical force law between the electrons and protons; the quantum ingredient turning the amorphous classical mess into an elaborate construction kit which enables one to understand the whole complicated edifice of the Periodic Table of chemical elements, the chemical interactions between atoms and molecules and also the cohesive forces which form them into solids, liquids and gases.

The allowed electron orbitals in an atom make interlocking patterns which, when complete, form a closed fuzzy ball of electric charge around the atomic nucleus. The noble gases are

* Our *h* is actually smaller by a factor of 2π than that originally defined by Planck.

those elements which have just enough electrons to complete one of these patterns. The electrical forces then tend to be completely saturated by the attraction to the nuclear protons, and such elements consequently have little tendency to combine with anything else.

The next most complicated atoms are the alkali metals. These have one electron orbiting outside the fuzzy ball which partially shields it from its own nucleus, so leaving it free to react with other atoms. It behaves like an electrical hook which latches particularly strongly into an element with one electron too few to complete the closed spherical structure, the extra electron in one element fitting comfortably into the slot produced by the missing electron in the other. The sodium chloride molecule is a typical example of this type of chemical bonding. This is a particularly simple chemical situation, but in principle one has in this purely electrical kit of parts the possibility of building up not only all chemical phenomena, but also the physical forces which hold molecules together in solids and liquids.

We have so far ignored gravitation. This requires explanation since there must be a gravitational attraction between the masses of the elementary particles, just as surely as there are electrical interactions between their charges. The strength of the gravitational interaction per unit mass can be found, for example, from the attraction of the Sun on the Earth. When this information is applied to the atom, it turns out that the gravitational attraction between a proton and an electron is weaker than the electrical attraction by an enormous factor of one thousand million million million million million million (10^{39})! This is one of the most remarkable pure numbers which turn up in the subject, but it quantifies a familiar situation. If a man jumps from a skyscraper window, the momentum he gains falling through a hundred metres, under the gravitational pull of the entire Earth, is lost in a fraction of a centimetre when he meets the cohesive electrical forces which form the hard surface of the sidewalk. Although he may be said to have been killed by his fall, it is the electrical forces which bring the fall to an abrupt end that actually do the damage. It is clear that in any chemical situation the effects of gravity can be safely neglected. But the atoms and molecules carry no net electric charge so, although they combine through forces which are purely electrical, they form familiar objects which, *in toto*, are electrically neutral. In normal life natural situations which are overtly electrical like thunderstorms are relatively rare. Hence it is the gravitational force, which was discovered first, of which we are most directly conscious in everyday life. Gravity keeps us on the surface of the Earth and maintains the Earth in its orbit round the Sun, so creating the exceptional environment which is essential to our existence.

So much for matter, but the synthesis covers an even wider range. Maxwell's equations, which were designed to summarise the behaviour of electric charges, have the remarkable consequence that an oscillating charge sets up a train of waves which can transmit energy and momentum through empty space without any accompanying transfer of mass. It was Maxwell's inspiration to identify this radiation with light which provides, in particular, the mechanism by which energy reaches the Earth from the Sun. Visible light is a very special form of electromagnetic radiation, and we have since become familiar, for example, with the long waves of radio communication and the short waves of radioactive fall out.

A wave is specified by the frequency of the oscillation, ω, and the wavelength which is the distance between successive wave crests. According to Maxwell's equations, light in empty space travels with a fixed velocity of 186 000 miles per second (3×10^8 metres per second). This means that short wavelengths are associated with high frequencies just as in a party of walkers, the one with the shortest stride must step fastest to keep pace with the rest. According to classical theory the energy in electromagnetic radiation can vary continuously and depends mainly on the intensity. But early this century several effects were found which required a different interpretation. Peculiar behaviour was discovered in the exchange of energy between atoms and radiation at some fixed temperature (black body radiation), in the energy balance when radiation is used to knock electrons out of a metal (photoelectric effect), and in the change in the wavelength of light when it reflects from an electron (Compton effect). All these effects can be explained quite simply on the basis of an idea first proposed by Planck that, at this atomic level, light behaves partly as a train of waves but also like a stream of particles, or quanta, called *photons*. These particles of light have energy and momentum, but no mass. (We came back to this below.) The energy of the photon determines the frequency of the corresponding wave, the constant which occurs in the relation between the two being Planck's constant h. $E = h\omega$. It was in this connection that Planck's constant was first introduced into physics in 1901. Note that this implies that the energy of a quantity of radiation of a given frequency, like that of an atom, can only change by discrete quantum jumps, corresponding to the addition or subtraction of single photons.

With the new quantum interpretation, Maxwell's equations not only determine the behaviour of free photons, but also describe how they interact with matter (that is with protons and electrons). It appears that, if we add photons to our list of elementary particles, the entire physical content of the Universe—both matter and radiation—can be explained in terms of these particles moving under the influence of electromagnetic and gravitational forces.

Statistical Effects

Before proceeding further, it should be stressed that this fundamental explanation, in principle, of physical effects in terms of protons, electrons and photons is not always carried out in practice. Physics would make very little progress, and would certainly have very few practical applications, if every situation had to be referred back to the primary laws of Newton and Maxwell. The standard technique for dealing with more complicated situations is to discover parameters which are adequate to describe the bulk behaviour of systems containing very large numbers of atoms and which can be understood, once and for all, in terms of the underlying fundamental forces. This latter analysis has often been worked out many years after the discovery of the useful parameters.

A very simple example is that of a perfect gas, which may be described in terms of its volume, the total heat, the temperature, and the pressure which it exerts on the walls of its container. The pressure, volume and temperature are simply related in a manner which has been known since the 17th century. In the fundamental theory the gas consists of a very diffuse swarm of a very large number of atoms in random motion. The heat is nothing but the total energy of motion of this swarm of atoms. The temperature is essentially the energy of a single typical atom. When two gases at different temperatures are mixed, the total heat, being the total energy, is just the sum of the heats of the two gases. In the subsequent collisions between the atoms the fast ones from the hot gas are slowed down, the slow ones from the cool gas are speeded up, till the whole system averages out to some temperature lying between the extremes of the original mixture.

The pressure is the cumulative effect of the atoms bouncing against the walls of the container which, for convenience, we take to be rectangular. It would take all the books of every library in the world to record the orbits of all the atoms, in terms in which one could forecast in detail the future behaviour of the gas. The information is completely superfluous. Just because there are so many atoms, it is safe to assume that, in effect, on average one-third of them are always moving parallel to each of the three directions defined by the edges of the container. The frequency with which the atoms bounce against the containing walls depends on how many there are, on the volume of the container and on the typical atomic speed. This speed is directly related to the temperature. A simple calculation expressing these facts explains why temperature, volume and pressure are related in the well-known way, and actually includes some extra information which relates these gross effects to the number of gas atoms to be found in a standard volume.

Another example is electric conductivity, which plays an essential role in even the simplest electric circuit theory. In a perfect crystalline solid, such as copper, the atomic nuclei lie in rows, like ranks of soldiers, to form a regular lattice. The electrons in the atoms are attracted to their own nuclei, but they are also pulled either way by the nuclei of all the other atoms. The net effect depends on the substance being considered, but a quantum mechanical calculation shows that for some solids, of which copper is typical, a proportion of the electrons are free to move through the lattice. These substances are good conductors, and in this idealised model, they are actually perfect conductors, the electrons flowing quite freely through the lattice like children running between the ranks of soldiers. However, in practice the lattice has imperfections—the ranks are not quite regular—and at room temperature the nuclei are oscillating about their mean positions, causing the electrons to collide with them, as the children might bump into the soldiers if they stepped from their strict positions in the ranks. This is the explanation of electrical conductivity, which can be summed up in a single number—the resistance. The detailed theory is quite complicated, but heat is also conducted through a metal by the same mechanism, so there are simple connections between electric and thermal conductivity which can be checked in some detail.

We will not now consider these pragmatic problems any further, but continue the discussion of the fundamental theory. The relation between the statistical effects of large numbers and the primary physical laws is considered again in Chapters 4 and 5 in connection with the direction of flow of time, and in the final chapter when we analyse the nature of physical prediction.

Conservation Laws

In the analysis of any situation it is always important to identify these features which remain constant against a general background of change and possible decay. In the study of a man's life an unswerving devotion to his mother or a single-minded pursuit of financial gain may provide a clue to a whole variety of actions, which at first appear disconnected and bizarre. In physics these constant guide lines are the conservation laws, which specify these properties of a closed system which remain the same, regardless of the motions of the constituent parts. At the end of the 19th century physical science recognised four such laws. These were:

Conservation of mass,
Conservation of energy,
Conservation of momentum,
Conservation of angular momentum.

The conservation of momentum is familiar to anyone who has failed to allow for it in jumping from a rowing-boat to the shore. In a free system, initially at rest, the momentum gained by one part in one direction has to be balanced by an equal momentum of the other parts in the opposite direction, the total adding up to zero. When a relatively heavy man steps from a light boat, his motion towards the bank must be balanced by a rapid recoil of the boat into midstream. If the man is unlucky the net effect is to put him into the water. The propulsion and steering of a rocket in empty space is achieved in just this way through the recoil of the main body of the spacecraft balancing the momentum of fuel ejected in the opposite direction. The motion of the centre of gravity of the complete system—spacecraft plus fuel—remains unchanged.

We have already met the concept of angular momentum in connection with the quantum conditions, and its conservation is just the generalisation of the above considerations to rotational motion. A flat object, like a wheel, rotating slowly has the same angular momentum as a rapidly spinning shaft of equal mass. This law is exploited by a skater who sets himself rotating relatively slowly, with a skate and arms held well out, and then converts this motion to a rapid spin by confining himself as near as possible to a single axis. This is a spectacular display of the conservation of angular momentum.

At first sight the conservation of energy appears more complicated. The simplest form of energy is the energy of motion, and when two billiard balls collide, not only the total momentum, but also the total kinetic energy is unchanged. However, when a fast car is brought screeching to a halt the kinetic energy of the motion is converted mainly into heat which goes, via the brake drums, into the atmosphere. But heat is just the kinetic energy of the atoms and molecules in the substance of the brake drums, which conveys itself by jostling to the molecules of the air. Thus this energy, too, is kinetic, but the motion is of constituents rather than the whole system. This is typical of the general situation. Some of the car's energy is also carried off as sound, which can be traced to the oscillations set up in the particles of the atmosphere. All this energy added up would be found to equal the original energy of the motion of the car.

The energy conveyed by radiation may also be regarded as due to the motion of the photons so that at the atomic level all the energy whether it be heat, sound, radiation or kinetic, may be regarded as due to the motion of the elementary particles.

The conservation of mass, looks, on the face of it, the most obvious law of all, since it appears to state that the total number of protons and electrons in the Universe is constant. In any closed system—which by definition is one in which no particles are being fed in or taken away—it is reasonable to suppose that the total number of protons and electrons remains constant, and this would appear to imply that the total mass remains unchanged. This law is valid to a very good approximation in normal circumstances, and is still used in chemistry. However, as a result of Maxwell's equations and the theory of relativity these last two conservation laws of mass and energy have undergone modifications with very far-reaching implications.

Relativity and the Conservation Laws

The velocity of light in empty space, determined by Maxwell's theory, is the same whatever the relative speed of the source and the receiver, regardless of the speed of either of these with respect to the Universe as a whole. This is quite different from the behaviour of either a bullet or a sound wave. A bullet fired in a fast-moving train at another passenger on the train will appear to behave normally to an observer on the train, but would hit someone standing near the track with the usual muzzle velocity modified by the velocity of the train. But light emitted from a moving source travels at the same speed as that from a stationary source. Einstein showed in the special theory of relativity that this peculiar behaviour of light implies some peculiar things about space and time, whose physical properties are defined by the measured distances and time intervals between events.

Let us consider a bank robbery in which a thief fires a revolver, and a clerk falls dead. From the point of view of a customer who observes the scene, the moving object is the bullet and the two events—the firing and the killing—are separated by some distance in space and some period in time. The bullet (or a fictitious observer we can imagine sitting on the bullet) sees it all quite differently. At the instant of the firing the thief rushes away from him and the victim towards him, the latter dropping dead at the moment when he rams himself into the bullet. According to the bullet, both events happen at the same place (that is where the bullet is) so that he would maintain that the distance between them was zero. However, on classical (and common sense) notions of time he would agree with the conventional observer about the time interval between the two events. Einstein showed that this is approximately true only if the bullet is moving slowly compared with the velocity of light. Actually the time interval would also appear shorter to the bullet, and the faster the bullet goes the shorter the interval would become. If the bullet were replaced by a laser beam (moving with the velocity of light) the time interval for an observer moving with the beam would also be reduced to zero. Thus for an observer moving with the laser beam the firing and the killing take place on the same spot and at the same instant. According to relativity theory no particle can travel faster than light, which is the equivalent of infinite velocity in a classical theory. This has the consequence that no observer can move so fast that he sees the clerk die *before* the shot is fired. Since our general objective is to find a quantative relation between cause and effect, this restriction is very important!

The theory of relativity implies that, in spite of these curious properties of space-time, genuine conservation laws must be valid for all observers. If energy is properly defined, both the customer and the fictitious observer sitting on the bullet must agree that the total energy is unchanged by the regrettable incident. To maintain the conservation laws under these extreme circumstances some modifications are necessary, particularly when speeds comparable with the velocity of light are involved. What actually happens is remarkable. The two classical conservation laws of mass and energy coalesce into a single law. This is because matter turns out to be just another form of energy. If c is the velocity of light, a particle of mass m has a certain, so-called, rest energy (because unlike kinetic energy it is still there when the particle is at rest), given by Einstein's well-known relation

$$E_{\text{rest}} = mc^2. \qquad (1)$$

If the particle is set in motion with a velocity v, small compared with c, its energy of motion is given approximately by the familiar formula

$$E_{\text{kinetic}} \simeq \tfrac{1}{2}mv^2. \qquad (2)$$

The total energy under these circumstances is the sum of these two

$$E \simeq mc^2 + \tfrac{1}{2}mv^2 \qquad (3)$$

The correct expression, valid for all allowed velocities, is more complicated;

$$E = \frac{mc^2}{\sqrt{1 - \frac{v^2}{c^2}}}. \qquad (4)$$

When the velocity is zero this clearly just gives the rest energy. For velocities v, which are small compared with c, it reduces approximately to the expression (3). Energy defined in this way is conserved for all observers. It has the special feature that for a massive particle the energy becomes infinite as the speed approaches that of light. Since one can only feed in a finite amount of energy, no massive particle ever actually reaches this velocity.

Because the velocity of light is so big by everyday standards ($3 \cdot 10^8$ metres per second), the rest energy is enormous. A two-ton spacecraft must be given a velocity of about 10 000 metres per second (25 000 miles per hour) in order that it can escape from the Earth. Even in these extreme circumstances the energy of motion is only equivalent to the rest energy of one-hundredth of a gramme.

In most physical situations the correctly defined total energy of a system can be expressed to a very good approximation as a sum of contributions, some very big ones coming from the rest energy, and some relatively tiny contributions from the energy of motion. When considering the over-all conservation of energy, the large rest energy contributions usually remain unchanged and when considered separately lead to the equivalent observation that the mass has remained constant throughout the process. This is the classical equation of 'conservation of mass'. The remaining small bits can then be equated to give the classical 'conservation of energy'. But there is really only one law, which is that the total energy is conserved. In a general situation which involves velocities comparable to the velocity of light the kinetic energy is comparable to the rest energy. In such cases no meaningful separation can be made and—most important of all—energy which appears as rest energy (mass) at the beginning of the process may be converted into kinetic energy (motion) by the end—and vice versa! When an atom emits a photon it is subsequently lighter by an amount just equivalent to the kinetic energy of the emitted radiation and the atomic recoil. Even in an exothermic chemical process the amount of material present at the end of a reaction is less (in mass) than at the beginning by an amount just equivalent to the energy released. The change in mass is so small that it is completely negligible when working to the accuracy of a measurement normally used by chemists, so the 'conservation of mass' is still a very good working rule in a chemical context.

The conservation laws of energy, momentum and angular momentum, as refined by the theory of relativity, were found by considering macroscopic matter, but allowing for velocities comparable to that of light. When applied to microscopic systems, such as atoms, it is found that they also apply to quantum mechanics without change. However, an electron whirling about a proton in, say, a hydrogen atom is found to behave rather like the Earth going round the Sun. In addition to its orbital motion it is spinning about its own axis, like a top. This spinning motion makes an additional contribution of $\frac{1}{2}h$ to the angular momentum, over and above that arising from the whirling motion. This is known as the spin. A proton also has spin of $\frac{1}{2}h$ and the photon a spin of h.

Matter and Anti-matter

Relativity, quantum mechanics and the generalised version of the energy conservation law were first fruitfully employed in conjunction by Dirac in 1928, when he developed the relativistic quantum theory of the electron. This theory is amazingly successful because it explained, in a natural way, why an electron has a spin of $\frac{1}{2}h$, and also correctly predicted its behaviour when subjected to a magnetic field. However, it suffered from an appalling flaw, since it seemed to imply the existence of particles of negative mass. These would move to the right when pushed to the left, and a vehicle made of such matter would accelerate when you put the brakes on. This is totally unacceptable. But Dirac realised that, by exploiting to the full the interchangeability of rest energy and kinetic energy, a more subtle interpretation of his theory was possible. This preserved the essential condition that all electrons have positive mass and, therefore, respond in a sensible way to applied forces, but corresponding to electrons of negative charge it implied the existence of anti-electrons of positive charge—called positrons. The argument leading to this conclusion was so general that it could be applied to any particles, the existence of positively charged protons, for example, implying the existence of negatively charged anti-protons. Anti-matter can be built up from anti-protons and positrons, just as matter is constructed from protons and electrons, and interacts in a similar manner with radiation. Nothing unusual happens provided the two types of matter are kept apart, and it is quite possible that there are galaxies of anti-matter existing in the Universe. However, if matter and anti-matter come together all hell breaks loose, since particle and anti-particle immediately annihilate each other in pairs, disintegrating into lighter particles with almost all the available energy turning into energy of motion. When an electron and positron are brought together they convert instantaneously into two or three photons—the entire rest energy (mass) turning into radiant heat. The conversion factor is enormous, as we have seen. If one could propel a rocket by this absolutely total combustion of matter and anti-matter, even used quite inefficiently, a few grammes of fuel would take a big spacecraft to the Moon. It is also possible to create matter out of kinetic energy by the reverse process. If two photons of sufficient energy collide (or a single photon strikes the electric field of an atom), the kinetic energy can turn into rest energy by the creation of an electron-positron pair. So much for the conservation of mass!

In the machinery of particle pair annihilation, Nature has invented a more drastic form of majority rule than has ever been attempted by any political regime. It is as though, in a two-party system, an election were conducted by political opponents throughout the population slaughtering each other in pairs, until only the actual majority of the major party were left alive with no dissenting voice. If, in the course of time, opponents arise, they are dealt with in the same manner. For this reason anti-matter is not normally found in surroundings predominantly occupied by matter. Any anti-particles, which do get created, very soon collide with particles and annihilate. However, very high energy protons are continually entering the Earth's atmosphere in cosmic rays. On striking the atmosphere these give rise to high energy photons, which in turn can create electron-positron pairs. The positrons may travel several centimetres before annihilating with electrons and an experimenter has a chance to detect them. The existence of positrons was established in this way soon after Dirac drew attention to the possibility.

The percipient reader may have realised already that this possibility of making rest energy out of kinetic energy makes nonsense of our notion of a final solution to the general problem of physics in terms of elementary particles and a naïve application of the Greek analytical approach. We come back again to this much later, and close the important subject of conservation laws with two disconnected remarks.

Massless Particles

Since we have seen that matter (mass) is energy, and radiation is also energy, it is energy which is the fundamental physical quantity of which mass and radiation—matter and light—are two manifestations. This is a somewhat abstract notion and it is convenient to continue to think in terms of particles. The proton and electron are plausible enough, but the photon of zero mass is a less familiar concept. It can be understood in the following way. The number *one* can be written in an infinite variety of ways including the following:

$$1 = 2 \times \frac{1}{2} = 3 \times \frac{1}{3} = 4 \times \frac{1}{4} = \ldots = n \times \frac{1}{n} = \ldots.$$

As we proceed the first factor gets bigger and the last smaller, so that ultimately we have infinity times zero, still equal to one. The momentum of a photon is essentially the product of its mass and its velocity. The velocity of a photon in a relativistic theory is equivalent to infinite velocity in a classical theory.

In this limiting sense, even if its mass is zero, the momentum of a photon can still be finite. A similar argument applies to the photon energy.

Invariance and Conservation Laws

The laws of energy and momentum conservation were first derived as a consequence of Newton's Laws and Maxwell's equations. It has since been realised that they follow from much more general considerations. We consider a closed system, like an isolated hydrogen molecule or the solar system, which moves only under the action of internal forces and is not subject to any appreciable external influence. It is reasonable to suppose that the behaviour of such a system is independent of its exact location in space, and that its time development is independent of the particular instant at which it is set in motion. Equivalently, we assume that identical experiments will give the same results whether they are set up in London or Moscow, in 1970 or 1984. The remarkable fact is that this assumed invariance of the behaviour of a closed system with respect to displacements in either time or space already implies the conservation of energy and momentum. In a similar manner the conservation of angular momentum follows from the assumption that the development of a closed system does not depend on its orientation. These conservation laws are thus not a special consequence of the theories of Newton and Maxwell, but would follow for any force laws which had these general invariance properties.

To get an idea of the significance of this, it is interesting to imagine how things might have been otherwise. On some theories the entire Universe is undergoing a vast series of rhythmic expansions and contractions like the heavy breathing of some enormous monster. It is conceivable that the development of the solar system could be related in some way to this cosmic pulsation. Then the subsequent development would depend, to some extent, on whether the formation of the planets took place during an expanding or contracting phase of the Universe. In this case the motion of the solar system would not be invariant for a displacement in time, since such a displacement would shift it to a different phase of the cosmic cycle. The general rule, then, says that this lack of invariance with respect to displacements in time would automatically imply that the energy of the solar system was not conserved, but that there was some exchange of energy between it and the cosmic breathing.

This relation between the invariance of physical systems with respect to some type of change and the conservation of some related quantity is a very general rule, which will play a crucial role in the subsequent discussion. We have stated it without proof, but it is suggested by our previous discussion of the solar system that this relation between invariance principles and conservation laws has an element of tautology about it. The freedom to transform a system in specified ways without altering its motion implies that certain influences, which could be present, are actually excluded. It is not altogether surprising that this in turn implies that some related property remains unchanged throughout the motion.

2

The Nucleus

The Atomic Nucleus, Mass and Size

The synthesis of physical phenomena in terms of electromagnetism and gravity, protons, electrons and photons was so fantastically successful and encompassed such a vast and diverse range of effects, that some quite level-headed physicists got carried away.

Planck's constant, h, the velocity of light, c, and the charge on the proton, e, are all physical constants in the sense that they stand, respectively, for a certain fixed quantum of angular momentum, a particular velocity and a certain amount of electric charge, each of which has a universal significance. They are particularly important because, in turn, they play a crucial role in quantum theory, in relativity and in electromagnetism. Their numerical values depend on the particular system of units which are used. Thus the velocity of light may variously be expressed as 186 000 miles per second or 3×10^8 metres per second. However, a certain special combination of all three of them, namely

$$\frac{e^2}{hc} = \frac{1}{137}$$

is a pure number, which is independent of the units used. It is known as the 'fine structure constant', because it first appeared in formulae specifying certain fine details in the allowed energy levels of hydrogen. It also determines the probability that an accelerating electron will emit or absorb a photon. It is a very convenient measure of the natural strength of electromagnetic phenomena. Another pure number which appears in the synthesis is the ratio of the masses of the proton and the electron.

Sir Arthur Eddington, who was certainly no fool, was so

overwhelmed by the success of the physics of his day that he spent the last years of his life trying to explain it away as a monstrous intellectual hoax, and attempted to establish his point by deriving these two pure numbers by pure thought. Apart from these two numbers, the subject appeared to be almost closed. Eddington produced the intellectually stimulating idea that the whole of physics might possibly be analogous to the activities of an ichthyologist (or specialist in small fish), who devoted a long and distinguished scientific career to the discovery of the basic biological law which bears his name. The law states that 'no fish are less than two inches long'. The ichthyologist never realised that his 'law' was nothing but a reflection of the size of the mesh of the net in which he caught the specimen for his study. This picturesque analogy emphasises the important point that the scientific approach is highly selective, but there seems no justification nowadays for the view that the whole content of physics is determined by the techniques which define its boundaries—that nothing comes out which has not implicitly been put in. Eddington had no reason to be embarrassed by the physicists' success. Fundamental physics is very far from being a closed subject and efforts to close it since 1930 have progressed rapidly backwards. This is the main burden of the rest of our tale.

We should stress again that there are many problems in physics which, although explicable in principle on the basis of primary laws, are difficult to interpret in practice. Outstanding examples until relatively recently were the extraordinary behaviour of helium when cooled to near absolute zero, when it becomes a 'superfluid', and the somewhat analogous electrical properties of superconductors. These are so remarkable that there was a possibility that they exhibited essentially new characteristics, which went beyond accepted physical concepts. But developments during the last fifteen years have established that even these lie within the established scope.

We are not interested here in such effects but in those, if any, which cannot be explained, even in principle, without introducing essentially new elements into the theory. Although it has been asserted that virtually everything can be explained by the above scheme, this turns out to be an overstatement.

The crucial domain is the nucleus of the atom itself, which has so far been treated rather uncritically as a conglomeration of enough protons to account for its electric charge. This is the only property which is necessary to explain the behaviour of the electrons in the atom. This, in turn, determines the interaction of atoms with radiation and their chemical characteristics. These problems almost totally preoccupied physicists up to the time we have considered, but an all-embracing scheme must also explain the internal structure of the nucleus; not only its electrical properties but its mass and the forces which hold it together. One of the simplifying features which made the development of a detailed atomic theory possible was that the nucleus could apparently be treated as inert lump, which supplied a central core of charge to the atom, but appeared to play no further part. This remarkable stability of the nucleus simplified the discussion of the atom, but now, itself, calls for an explanation.

The typical atom has a radius of about 10^{-10} metres. This can be found roughly but readily from the observation that a teaspoonful of oil, given time, will spread over the whole surface of a fair-sized swimming-pool. From the volume of oil involved and the area of the pool one can get a measure of the thickness of the slick which must be at least a mono-molecular layer. In this way one obtains a simple upper limit to atomic size.

The size of the nucleus was deduced by Rutherford in his famous experiment carried out in Manchester in 1911. This is the prototype of all nuclear experiments. Rutherford was working on radioactive nuclei, some of which emit high energy α-particles. These are just helium nuclei. A collimated beam of these particles was directed at a thin gold foil, and their deflections were measured by observing the tiny flashes which they made on hitting a fluorescent screen placed around it. The flashes were just visible to the naked eye, if the observer first

I

A view of the CERN laboratory just outside Geneva, showing the 28 GeV proton accelerator ring under its mound of earth and the experimental halls. Also visible, coming away tangentially from the machine ring, is a neutrino beam line leading to a bubble chamber facility. *(Photo: CERN)*

A view of the interior of the machine ring of the CERN 28 GeV proton accelerator, which lies buried under the circular mound visible in Plate I. In the left foreground are two of the hundred 34-ton magnets which keep the protons on their circular track of 100 metre radius. The cylindrical vacuum pipe in which the protons travel can be seen entering the gap between the poles of the C-shaped magnet at the left of the picture. *(Photo: CERN)*

spent a preparatory period of twenty minutes in total darkness.

The α-particles have twice the charge and about 8000 times the mass of an electron. Thus they simply brush aside the electrons in the gold foil like cannon balls passing through a swarm of bees. Since gravitational forces have been seen to be quite negligible, the only thing which could cause the α-particles to be deflected from the collimated path would be the electric repulsion between them and the heavy gold nuclei, containing a charge of seventy-nine protons. However, it was generally believed at that time that this charge in a heavy nucleus was fairly widely and evenly distributed in a volume only a bit smaller than the atom itself. It was, after all, sensible to assume that solid matter was reasonably solid! On this basis the α-particle as it passed through the gold would go through a layer of charge carried by the gold nuclei which was evenly distributed in all directions. The electric repulsions between the α-particle and the gold nuclei would tend to cancel each other out. The α-particle would be pushed and pulled equally in all directions and consequently there would be no reason to expect it to deviate appreciably from its original path. Although the vast majority of α-particles went almost straight through the gold foil, to Rutherford's amazement, a number were deflected through 5° or 10°, and some even failed to penetrate the foil, bouncing back on the side from which they originally entered. It was as though a squadron of light tanks had been ordered to advance in pitch darkness through what had been taken for an easily penetrable hedge and had found themselves, instead, crossing an avenue of widely-spaced but well-developed trees. Those vehicles which went through the gaps would not notice the difference, but the ones which hit the trees would either be knocked off course, or be stopped in their tracks by head-on collisions. In exactly the same way, if it is assumed that the charge in the gold nuclei is not spread uniformly through the foil, but is in fact separated into very highly concentrated cores, it is possible for an occasional α-particle to pass very close to one nucleus and relatively far from all the others. When the distance between two charges is halved the force between them is quadrupled; one-tenth the distance, one hundred times the force; and so on. The force between charges builds up very rapidly as the distance between them is reduced. A close shave past a highly concentrated gold nucleus, like a glancing collision with a tree in the avenue, could produce a very strong electrical repulsion, not balanced by any other forces, which would appreciably change the α-particle's direction. The more concentrated the charge the larger the possible effect, so the largest observed deflections give a measure of the nuclear radius. Rutherford came to the astounding conclusion that the nucleus has a radius about ten thousand times smaller than that of the atom. If an atom were drawn in a scale such that the electron orbits are comparable with those of the planets, the nucleus would still be much smaller than the sun. Solid matter, in terms of its mass distribution, turns out to be almost entirely empty space—even emptier than the solar system. If it were not for the long-range electrical forces, which give them both cohesion, a man could go straight through a barn door like a few flies passing through very wide mesh wire netting.

So much for the size, but the masses of the nuclei also presented a problem. Except for ordinary hydrogen these masses are found to be roughly double what one would expect if the nuclei consisted solely of enough protons to make up the required electric charge. This problem was not solved until 1932. It was then discovered that when a beryllium nucleus is bombarded with α-particles, as in the Rutherford experiment, the α-particles and beryllium combine to form a single heavier nucleus containing all the available charge, and the surplus energy is carried off by a neutral particle. At first this was thought to be a photon, but it was then found that in subsequent collisions this neutral particle could give up nearly all its energy to a proton. When a stationary billiard ball is struck by something relatively light like a table-tennis ball, the lighter particle tends to retain the motion. On the other hand, if the stationary billiard ball is struck, head on, by another of equal mass the motion swaps over from one ball to the other. These different effects follow from the conservation of energy and momentum. By such arguments it was shown that the observations could only be explained on the assumption that the neutral particles were almost equal in mass to the protons. They are called *neutrons* and must be added to the list of elementary particles. It is consistent with all observed nuclear reactions to suppose that a nucleus is made up of neutrons and protons—enough protons to give the required electric charge and then a roughly equal number of neutrons to make up the mass.

The Nuclear Force

We must now consider the problem of the amazing stability of this nuclear structure. This stability is made all the more curious by the smallness of the Rutherford radius. It is immediately clear that it is totally inexplicable on the basis of electromagnetism and gravity. If the distance between two protons in a nucleus is ten thousand (10^4) times smaller than that between an electron and a proton in an atom, by the rule stated above, the electrical force between them is a hundred million times (10^8) bigger. But it is working in the wrong direction. Since all protons in the nucleus carry a like positive charge, this enormous force is a repulsion tending to blow the nucleus apart. The only other resource available from classical physics is gravity, which does at least work in the right direction. However, it is weaker than the electrical repulsion by the enormous factor similar to the one already quoted, in connection with atomic structure (10^{36}). It is true that, because of the extra proton mass, the gravitational attraction between two protons is a thousand times bigger than that between an electron and a proton, while the magnitude of the electrical force is the same. However, among all those millions a factor of a thousand makes no difference to the qualitative result. Gravitational forces are overwhelmed by electrical forces inside a nucleus, just as effectively as they are in an atom. The only conclusion one can draw is that at the tiny distances inside atomic nuclei of 10^{-14} metres some completely new, specifically nuclear, force is operating, strong enough to overcome the enormous electrical forces which build up at these distances. Something new and enormously powerful is going on here which lies completely outside the classical scheme. In this very important respect nuclear physics and its offspring, sub-nuclear (or high energy) physics, are unique. For the last forty years the problem of expanding the domain of validity of fundamental physics—the search for new physical laws of the same basic generality as those of Newton and Maxwell—has lain in the unravelling of nuclear and sub-nuclear phenomena. There has been intense activity in this field, particularly since the Second World War, and we have had an experience which can be compared to that of Schliemann when, in his search for Troy, he found not one but nine cities buried one below the other in the great mound of Hissarlik. After working through the layer of the chemical elements to the atoms, it appeared that one had reached bedrock. The atomic nucleus disturbed this complacent conclusion like the finding of a single shard of pottery which did not fit into the context of the civilisations already discovered and, stranger still, by the hard brightness of its gloss gave evidence of a knowledge of powerful techniques completely alien to those implied by the rest of its surroundings. The study of this single clue has led us down to new layers of reality, whose existence was not even dreamed of thirty years ago. Further clues have been found which suggest even deeper layers and there is no knowing when we shall again feel that we are standing on ground so firm that further exploration is superfluous.

3

Strong and Weak

Particle Accelerators and Detectors

The approach to the study of the nucleus is still basically the analytical one of Democritus. The objective is to take it to pieces, or to probe it with some yet smaller structure. In the study of the atom the appropriate probe was electromagnetic radiation. Electrons can actually be knocked out of an atom by hitting it with sufficiently energetic photons. When subjected to gentler blows the electron can be made to jump from one allowed energy level to another. In this way a systematic pattern can be built up of all the possible configurations, and hence a picture is formed of the entire structure. The most useful information has come from very detailed studies of the hydrogen atom, because it is so simple, and consequently information is relatively easy to interpret. The energy levels of hydrogen can now be calculated, in agreement with experiment, to an accuracy of one part in a million. These calculations include very subtle effects due to the magnetic field set up by the proton spin and the back-reaction of the radiation emitted by the orbiting electron. In general the method works because an atom is electrical in nature. The photon interacts with an electrical structure and is itself much simpler than the atom which is being investigated. The natural tool-kit for dismantling an atom is a set of photons of various energies.

The situation in nuclear physics is quite different. The objective is to study the nuclear particles, proton and neutron (which we shall call nucleons), and the strong nuclear force which operates between them. In a typical nucleus with anything from four to over two hundred nucleons the situation is far too complicated to be sorted out in any detail, and the nuclear analogue of the hydrogen atom is a system consisting of just two nucleons. But what can one use for a probe? In this context the photon is not directly relevant because we are not primarily interested in the electrical structure (although recently experiments with photons and electrons have given interesting subsidiary information). Essentially the only experiment one can perform is to bang the two nucleons together and see what happens. This is rather crude. It is like setting out to learn about the internal-combustion engine by studying the accident reports of collisions between cars on the highway. But it is the best that we can do, and in this respect the famous Rutherford experiment, which consisted of an elementary study of collisions between helium and gold nuclei, is typical of all experiments in this field. Rutherford's experimental technique was primitive by modern standards, because his beam of nuclear projectiles was restricted to what could be conveniently obtained from a natural radioactive source. Consequently, the energy was quite low, and the nuclei which suffered near collisions were pushed apart by the electrical repulsion between their charges before the nuclear force had a chance to operate. The information about nuclear forces from this experiment came indirectly from the deduced nuclear radius and the known stability of a nucleus in any chemical context.

To get any further it is necessary to work with protons at much higher energies, so that when fired at nuclear targets—ideally other protons in the form of liquid hydrogen—they are able to overcome the electrical repulsions, and penetrate to distances at which the specific nuclear forces are operating. To do this it is necessary to accelerate the protons artificially in the laboratory. This involves a major step forward in technique which enables one to increase not only the energy involved in collisions, but also the rate at which interesting collisions occur. For this reason the basic experimental tool of the subject is the proton accelerator.

We have already seen that a two-ton spacecraft at escape velocity has a kinetic energy equivalent to the rest energy of a tiny fraction of a gramme. In atomic physics the typical kinetic energies are also very very small compared to the rest energies of the particle involved. In nuclear physics, for the first time, the two types of energy—kinetic coming from the motion and rest energy due to the mass of the relevant particles—are always comparable, and the natural energy unit is the rest energy of a proton. This is usually quoted in giga (10^9) electron volts—abbreviated to GeV—one GeV being approximately equivalent to the rest energy of one nucleon. The two largest machines in operation during the 1960s were at CERN (Geneva) and Brookhaven (Long Island, USA) and both accelerated protons to a kinetic energy equal to about thirty times their rest energy or 30 GeV. (See Plate I.) More recently a 70 GeV machine has come into operation at the Serpukhov laboratory just outside Moscow. Accelerators which operate in the several hundred GeV region are to be commissioned during the 1970s. (See Table 1.)

In the course of the last twenty years there have been important refinements in the technique of machine construction, but basically all these accelerators consist of a circular ring of magnets, which keep a bunch of about ten million million (10^{13}) protons circulating in a vacuum chamber with cross-section about 10 cm $\times$ 6 cm. (See Plate II.) As the protons go round they receive a rhythmic series of accelerating electrical pulses which brings them, step by step, to their maximum energy, at which point they are deflected into the nuclear target. In a 30 GeV machine the protons make three hundred thousand circuits and the whole process of accelerating a bunch of protons is repeated about every three seconds. As the energy goes up so does the size of the machine, which however has to be built with very high precision. The large machine now coming into commission at Batavia in the USA is designed to accelerate protons to a kinetic energy four hundred (possibly later to nine hundred) times their own rest mass. The ring has a circumference of seven and a half kilometres (about five miles). The magnets weigh a total of 25 000 tons and each one has to be located with

TABLE 1

High Energy Proton Accelerators

A world list of proton accelerators which have been built to produce maximum kinetic energies in the GeV range. The rest energy of a proton is about 1 GeV. The single entry which lies below 1 GeV is a historic machine, since it was the first to be built capable of producing pions in proton-proton collisions. It was followed within a few years by five other similar machines in the USA and one at the University of Liverpool (UK) which first operated in 1955.

Year of commission	*Energy* (GeV)	*Radius metres*	*Laboratory*
1948	0·34		Berkeley (USA)
1952	3	10	Brookhaven (USA)
1953	1	4	Birmingham (UK)
1954	6·4	18	Berkeley (USA)
1957	10	30	Dubna (USSR)
1958	2·5	11	Saclay (France)
1961	28	100	CERN (Geneva)
1961	33	100	Brookhaven (USA)
1962	7	23	Rutherford (UK)
1962	3	12	Princeton (USA)
1962	12·5	27	Argonne (USA)
1968	70	236	Serpukhov (USSR)
(1972)	400	1000	Batavia (USA)

an accuracy of $\frac{1}{2}$ mm. The surveying system used when the magnet ring was laid for the 28 GeV machine at CERN (Geneva) was so accurate that it was sensitive to the pressure of the Atlantic tides coming into the Bay of Biscay three hundred miles away. The accelerating radio frequency field has to be kept exactly synchronised with the bunch of protons as they speed up, and the strength of the magnetic field must also be modified to confine them to the fixed circular race track. Relativistic effects are essential to the design. The electricity consumption of a 30 GeV machine would satisfy the needs of a town of a hundred thousand inhabitants. It is evident from Table 1 that there has been a steady increase in the maximum energy of proton accelerators built since the Second World War. These machines are to nuclear physics what the telescope is to astronomy. Each new increase in available proton energy has led to new discoveries in nuclear and sub-nuclear physics, just as larger telescopes and detectors of new wave-bands of radiation, such as radio telescopes, have led to new developments in the study of the Universe.

Comparable advances have been made in techniques for studying the after-effects of nuclear collisions. Rutherford's detection device consisted of an eager research student who was kept for twenty minutes in total darkness, after which he was able to detect the tiny flashes made by the deflected α-particles on a phosphorescent screen. Several significant events took place per minute. Charged particles emerging from nuclear collisions can now be detected with electronic devices and counted at the rate of a hundred million (10^8) per second. In favourable circumstances this information can be fed, on line, to a computer for analysis without human intervention. Such an approach is possible when one has already a clear picture of the situation and knows exactly the question which has to be answered. (See Plate VI(a).)

Alternatively, there are more generally exploratory techniques in which tracks of charged particles taking part in nuclear collisions are photographed. It is possible to record nuclear events which actually take place in specially prepared photographic emulsion. But most work of this type since 1950 has been done using the hydrogen bubble chamber which consists essentially of a bath of liquid hydrogen. (See Plate VI(b).) By suddenly reducing the pressure, this can be put in a condition in which it ought to be boiling, but has not had time to do so. In this highly unstable state, which lasts a few thousandths of a second, a very small trigger can set off an appreciable effect. A single charged particle, such as an electron or proton, passing through the liquid sets up a train of bubbles along its track which can be photographed. The nuclei of the hydrogen atoms in the liquid of the chamber constitute a very large number of stationary target protons. If a beam of protons, or other charged particles, from an accelerator is directed into the bubble chamber, a complete photographic record can be obtained of the tracks of both the incoming beam and the recoiling particles which result from any nuclear collisions which take place in the liquid. If the whole business is carried out in a strong magnetic field the curvature of the tracks gives additional information on the particle momenta. By combining stereoscopic views, it is possible to reconstruct the event in three dimensions. This analysis is also done nowadays with extensive use of computers, and a single experiment may involve the study of one million photographs. This technique is discussed in more detail below and a number of bubble chamber photographs are reproduced in Plates VIII–XII.

A third technique which combines some of the main advantages of both counters and bubble chambers is the spark chamber, which consists of layers of parallel metal plates separated by a noble gas. The system has the advantage over the bubble chamber that it can be triggered to operate under selected conditions. At the selected opportune moment a very large field is put across the gaps between the plates for a few micro-seconds. In these circumstances the spark discharged between the plates follows the line of a charged particle passing through the apparatus, giving a track which is less detailed than that of a bubble chamber, but contains considerably more information than a simple counter. (See Plate VII.)

The Strong Interaction and Hadrons

The picture of the nuclear and sub-nuclear world which has emerged from experiments performed with such equipment is surprising, and we are now in a position to spell out its main features. We have already compared a nuclear experiment to a collision between cars. The analogy is useful and we will continue to refer to it. A typical experiment in which a beam of protons from an accelerator is directed into a hydrogen target may be visualised as a stream of cars hurtling full throttle into a full car park. In liquid hydrogen the atoms are packed close together, but as we have seen, this still leaves the atomic nuclei—protons in this case—very far apart. Thus one should visualise the parked cars as widely spaced, and there is an appreciable chance of a car from the stream passing clean through the park without hitting anything. It is, therefore, reasonable to analyse what is seen on the assumption that each car is involved in at most one collision.

If the cars have drivers they will attempt to avoid collisions. This effect may be taken as the analogue of the electrical repulsion between the protons which at low energies can keep them apart. As the speed of the incoming cars increases, the drivers become less and less effective. Similarly, as the energy of the protons is raised, the influence of the electrical forces in avoiding collisions is reduced, and ultimately can be neglected. This is the situation in which we are interested.

The first thing one would notice under these circumstances is that a miss is as good as a mile. Provided the cars pass with their centres more than about six feet apart, nothing happens. The moment this distance is reduced, they bump into each other and very violent actions take place. There is operating between the cars what is described as a strong, short-range force. Outside the range there are no effects, but once you get inside the range you must expect trouble. Nucleon-nucleon collisions are found to behave in just this way. Clearly, the frequency of collisions depends on the size of the cars, the spacing between cars in the park and the intensity (number and speed) of the cars in the stream. Since the frequency of collisions can be observed and the spacing and intensity are known, the size of a car could be deduced from this information, if it were not already known. In exactly the same way one finds that the effective range of the nuclear force is about one order of magnitude smaller than the radius of a typical medium-sized nucleus. That is about one thousand million millionth of a metre (10^{-15} metres). We shall refer to this as the nucleon radius. If the distance between two protons is appreciably greater than this radius, nuclear forces between them are negligible, and the situation is dominated by electromagnetic effects—the drivers are in control—but once inside this range the nuclear forces take over. The strength of the electromagnetic forces are given by the fine structure constant $e^2/hc = 1/137$. The equivalent pure number which measures the strong nuclear effects has a value of about unity. This merely states that in any physical situation, like the inside of a nucleus, in which electromagnetic and strong nuclear effects are both operative, the latter will swamp the former by a factor of about a hundred. Under such circumstances it is quite a reasonable approximation to ignore electromagnetic effects altogether. Once the cars have collided the drivers can exert only a negligible influence over the situation.

In 1934 the Japanese physicist Yukawa suggested the mechanism through which this strong nuclear interaction operates. He proposed that a nucleon should not be thought of as a structureless point, but as a small blob of sub-nuclear matter, *Urmaterie*, which for want of a better word we shall call 'goo' The radius of the blob is equal to the nucleon radius and—speaking colloquially—it is very sticky. Provided the nucleons pass each other without touching, nothing happens. But once the two blobs of goo come in contact they coalesce instantaneously, and very strong interactions are set up. Their momentum tears the nucleons apart again, but their directions will have been modified, and they may be thought of as interacting through the exchange of goo while in contact.

This idea has a variety of consequences. In the first place it implies that the individual nucleons in a nucleus will pack together like tennis balls in a string bag with a density determined by the nucleon radius. Each one interacts only with those which are in immediate contact with it. This is to be compared with the electrons in an atom, which are more like a few flies buzzing about a large room, each one of which, however, interacts with all the others through the long-range electromagnetic interaction.

The Yukawa model also implies that the electric charge of the proton, which is carried by the goo, will not be concentrated at a point but spread over a region of dimensions given by the nucleon radius. Since the proton is spinning, this charge will form small current loops which give it magnetic properties. On the same basis the neutron may be an equal mixture of positively and negatively charged goo, which gives it magnetic properties, although its total electric charge is zero. These predictions have been dramatically confirmed during the last ten years in a study of deflections produced in a beam of very high energy electrons by target protons. The principle of these experiments is exactly similar to that of Rutherford, except that he thought he was aiming his charged projectiles at a fairly large target nucleus, and found from the observed large deflections, that the nucleus was much smaller than he anticipated. In the modern experiments a point proton would be the most efficient deflector of electrons imaginable. Its finite size can be deduced from a reduction in the number of observed deflections through large angles, compared to what would be expected from a point source. The results are in detailed agreement with Yukawa's hypothesis.

But Yukawa's proposal has even more startling implications. The blobs of goo are quantum mechanical systems in the sense that their angular momentum is comparable with h. In just the same way that this leads to atoms with a discrete set of allowed energy levels, it ensures that the blobs of goo are not able to exist with an arbitrary continuous range of masses, but with masses belonging to a discrete set of allowed values. The smallest is known as a π-meson—or pion—and we denote its mass by m_π. A glancing collision between two protons is controlled geometrically by the nucleon radius r_n, but on Yukawa's model it takes place, mechanically, through the exchange of the minimum allowed mass of goo, which is m_π. Consequently the nucleon radius and the pion mass must be related. Since we are dealing with a relativistic quantum mechanical situation c and h may be expected to occur in the relation. Now the nature of the physical properties represented by these symbols is as follows:

$$\begin{aligned} h &= \text{mass} \times \text{radius} \times \text{velocity}, \\ c &= \text{velocity}, \\ r_n &= \text{radius}, \\ m_\pi &= \text{mass}. \end{aligned}$$

Any sensible equation must connect physical quantities of the same type. From these very general considerations, it is clear that the four quantities above can only be related in one way, namely

$$m_\pi = \frac{h}{r_n c}$$

which is of the correct form:

$$\text{mass} = \frac{\text{mass} \times \text{radius} \times \text{velocity}}{\text{radius} \times \text{velocity}}.$$

Strictly speaking this argument is completely empty, because we could multiply the right-hand side of the equation by any pure number and it would still be dimensionally valid. However, experience has shown that if numbers very different from unity appear in equations in physics, they can usually be traced to some physical origin. For example, if electromagnetism were involved we would expect to meet numerical factors e^2/hc which would drastically affect the situation. However, we are considering a system dominated by the strong nuclear interaction for which the corresponding strength constant is unity. It is thus at least a reasonable guess to take the above equation, as it stands, at face value. Since h and e, and r_n are all known, this gives an estimate of m_π which works out at about one-seventh of the nucleon mass. The very important point is that, if the original nucleon-nucleon collision takes place with sufficient kinetic energy, there is the possibility that part of this will convert into rest energy and the final state after the collision will consist of two nucleons and a pion. This is the really crucial qualitative predictions of Yukawa's theory. Pions should be produced in nucleon-nucleon collisions.

If we can forget for the moment colliding cars and blobs of goo, the nucleons in the Yukawa mechanism may alternatively be thought of as football players, who interact when they are close enough to do so by the exchange of a ball, each one changing direction as he throws or catches his pass. The range of such an interaction clearly depends on the mass of the ball.

The players can exchange tennis balls over considerable distances, but must be almost in contact before they swap cannon balls. (This is just what our equation says. Since h and c are fixed as m_π goes up r_n must come down.) In the language of this analogy, the important prediction is that if the players are running fast enough a pass may be dropped and the loose ball—that is a pion—as well as the two players may emerge from the mêlée.

Experiments to check these ideas did not have to wait for the construction of sufficiently high energy proton accelerators. A fine rain of high energy particles—mostly protons—is continually falling on the Earth from outer space. Evidence for the existence of a particle with mass roughly equal to that produced for the pion began to accumulate just before the Second World War, from the study of the collision of these high energy cosmic rays with the particles in the Earth's atmosphere. Then came a break of about six years in which no work of this fundamental nature was done. When experiments were resumed it was found, very curiously, that the particle produced by cosmic rays, with mass about one-tenth that of the proton—called muon—does not interact strongly with nuclear matter. It cannot be identified with Yukawa's pion. The prediction that the pion is produced in nucleon-nucleon collisions implies that the reverse process must also be possible. On entering a nucleus a pion should be quickly absorbed with a considerable explosion, its rest energy converting into the kinetic energy of the absorbing nucleons. The muon does no such thing. It passes through several feet of steel without anything happening at all. This was cleared up by the Bristol group during 1947 using nuclear photographic emulsion exposed to cosmic rays. They showed that there are, in fact, two particles. There is a pion which has just the properties predicted by Yukawa, which disintegrates in about one hundred millionth of a second (10^{-8}) into a muon, μ, and a neutrino, ν_μ:

$$\pi^+ \longrightarrow \mu^+ + \nu_\mu{}^\circ.$$

The neutrino is a neutral massless particle which differs from the photon in that it has spin $\frac{1}{2}h$, rather than spin $1h$. This curious complication was not foreseen, although in character it was not completely new. It was already known that a free neutron disintegrates in a similar manner into a proton, an electron and a neutrino, ν_e,

$$n^\circ \longrightarrow p^+ + e^- + \bar{\nu}_e^\circ.$$

This process was first observed about 1930 for neutrons inside nuclei. Since the high energy rays emitted by the nucleus were dubbed β-radiation before they were correctly identified as electrons, the effect is still known as β-decay.

We return to these effects later. Let us get back to the main argument. The general picture, which had emerged from exploratory work in cosmic rays, was confirmed in detail when proton accelerators with sufficient energy to produce pions came into operation in the USA, and the development of the hydrogen bubble chamber made possible the systematic study of large numbers of photographs of elementary particle collisions. Particularly important in this connection was the commissioning of the Berkeley accelerator in 1948. Yukawa's proposal was triumphantly and conclusively vindicated in all its qualitative aspects in experiments made possible with this machine.

What was not anticipated was that the pion and nucleon would prove to be only the least massive states of a whole spectrum of blobs of sub-nuclear goo. The first indications of this also came in 1947 from experiments in cosmic rays, but the subject was soon dominated by the much more systematic experiments which are possible with proton accelerators and hydrogen bubble chambers. The subsequent ten years saw a very rapid development in this field leading to the situation which is summarised in Table 2. The most remarkable thing about this table is just that it exists and is so long and complicated. Nuclear physics was undertaken as an investigation of the bedrock of the physical world, and everything should have been simple and straightforward. Clearly just the reverse is the case.

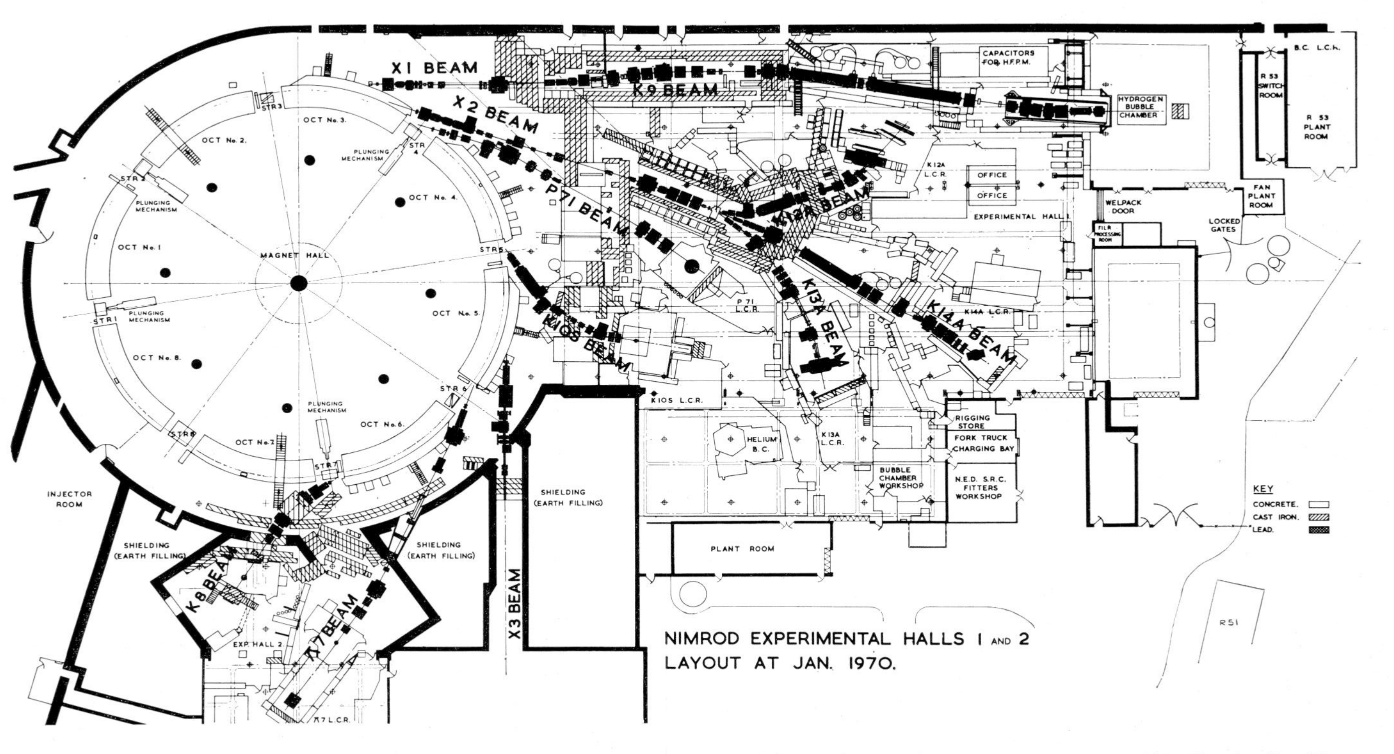

A ground plan of the 7 GeV proton accelerator NIMROD at the Rutherford Laboratory, UK, showing the machine ring (radius 23 metres) and various particle beam lines leading to bubble chamber and counter apparatus of the type shown in Plates VI (a) and (b) the experimental halls.
(*Photo: Rutherford Laboratory*)

A view of the NIMROD 7 GeV magnet ring and, coming tangentially from it, a beam line of extracted particles as shown in the plan on Plate III. The vacuum pipe of the beam line and a complicated series of bending and

Let us consider the information contained in this table first and then discuss the experiments through which it was discovered.

The proton and neutron are the two lightest members of a whole family of baryons (heavy ones) of spin $\frac{1}{2}h$, which appear in multiplets with one, two or three members, with very similar masses, but different electric charges. The observed charges are all small integer multiples of the charge on the electron. The heavier baryons are denoted by capital Greek letters Ξ, Σ and Λ.

Yukawa's π-meson, or pion, is the lightest member of another whole family of particles with no spin which are known collectively as mesons (the middle ones). They also appear in multiplets of similar mass but different charge and the heavier ones are distinguished individually by symbols of K and η.

All these particles are copiously produced in nuclear collisions provided only that enough kinetic energy is available to convert into the required extra rest mass. Corresponding to each particle there is an anti-particle, which can also be produced, subject to conditions which are discussed in more detail below. After production, the anti-particles lead normal lives until they come in contact with particles, when the goo and anti-goo annihilates into a great blast of kinetic energy. A proton and anti-proton, for example, annihilate each other, leaving behind a few mesons, 80% of the original rest mass converting into kinetic energy. (See Plate IX.) All these effects are consequences of the strong nuclear interaction which holds the atomic nuclei together as stable units throughout all chemical processes. The particles which interact through this strong force are known as the strongly interacting particles—or *hadrons*. This is complicated enough, but there are other things going on as well.

The Weak Interaction and Leptons

The proton is unique among the hadrons in being a stable particle. Left to itself a proton, as far as we know, would last through all eternity. (The basis for this belief is discussed more fully below.) All the other hadrons have the property, already discovered for the neutron and pion, that even if left entirely undisturbed they disintegrate—or decay—into lighter particles. The various main decay schemes are also indicated in Table 2, and the mean life for the hadrons is seen to be about one ten thousand millionth of a second (10^{-10} seconds). This time between birth and death of the particle is that measured by a clock travelling with the particle. As we have seen above, when discussing the bullet and the bank clerk, to the experimenter who

TABLE 2

The sub-nuclear particles more or less as they were known around 1957, giving the mass, spin, mean life time and the decay products of their spontaneous disintegration. The upper suffix on the particle labels denotes the electric charge. The hadrons interact with each other through the Strong interaction. The photon interacts with all charged particles through the electromagnetic interaction. The leptons have only electromagnetic and Weak interactions.

		Particle	*Mass* (GeV/c^2)	*Mean life seconds*	*Decay products*	*Anti-particle*
HADRONS				*Baryons (spin ½ h)*		
		Ξ^-	1·321	10^{-10}	$\rightarrow \Lambda^\circ + \pi^-$	$\bar{\Xi}^+$
		Ξ°	1·314	10^{-10}	$\rightarrow \Lambda^\circ + \pi^\circ$	$\bar{\Xi}^\circ$
		Σ^-	1·197	10^{-10}	$\rightarrow n^\circ + \pi^-$	$\bar{\Sigma}^+$
		Σ°	1·192	10^{-18}	$\rightarrow \Lambda^\circ + \gamma$	$\bar{\Sigma}^\circ$
		Σ^+	1·189	10^{-10}	$\rightarrow p^+ + \pi^\circ, n^\circ + \pi^+$	$\bar{\Sigma}^-$
		Λ°	1·115	10^{-10}	$\rightarrow p^+ + \pi^-$	$\bar{\Lambda}^\circ$
	neutron	n°	0·9395	$3 \times 10^{+3}$	$\rightarrow p^+ + e^- + \nu_e^\circ$	$\bar{n}^\circ$
	proton	p^+	0·9382	stable		$\bar{p}^-$
				Mesons (Spin 0)		
		K^+	0·494	10^{-8}	$\rightarrow \pi^+ + \pi^\circ$	$\bar{K}^-$
		K°	0·498	10^{-10}	$\rightarrow \pi^+ + \pi^-$	$\bar{K}^\circ$
				10^{-8}	$\rightarrow \pi^+ + \pi^- + \pi^\circ$	
	pion	π^+	0·139	10^{-8}	$\rightarrow \mu^+ + \nu_\mu^\circ, e^+ + \nu_e^\circ$	π^-
		π°	0·134	10^{-16}	$\rightarrow 2\gamma$	π°
		η°	0·549	10^{-18}	$\rightarrow 2\gamma$	η°
	photon	γ	0	stable		
LEPTONS	muon	μ^-	0·105	10^{-6}	$\rightarrow e^- + \bar{\nu}_e^\circ + \nu_\mu^\circ$	$\bar{\mu}^+$
		ν_e°	0	stable		$\bar{\nu}_e^\circ$
	neutrino	ν_μ°	0	stable		$\bar{\nu}_\mu^\circ$
	electron	e^-	0·0005	stable	(positron)	e^+

watches the particle go past with velocity approaching the speed of light, it appears to live much longer. In this time it can cover several centimetres which is long enough for the experimenter to establish its existence, for example, by a track in a bubble chamber. Even so, by normal standards, these particles live for a very short time. The crucial thing here is that from the point of view of the nucleon the time is enormously long!

We already know the typical nucleon radius

$$r_n = \frac{h}{m_\pi c} \simeq 10^{-15} \text{ metres.}$$

To construct the typical nuclear time, we must divide this by the velocity relevant to the system. This can only be the velocity of light. The typical nuclear time is the interval required for a light flash to cover a distance equal to the nucleon radius. Let us call this the nuclear year and denote it by τ_n. Then

$$\tau_n = \frac{r_n}{c} = \frac{h}{m_\pi c^2} \simeq 10^{-23} \text{ seconds.}$$

The nuclear year, which is the natural time scale of sub-nuclear phenomena, is about one million million million millionth of a second. If the disruptive forces which caused the hadrons to disintegrate were the same strong forces through which they are created, their mean lives would have to be of this order of magnitude.

In fact they are observed to live for about ten million million (10^{13}) nuclear years. This is remarkable. A number of this magnitude implies that yet again something completely new is going on, which is totally inexplicable on the basis of what we have assumed so far. In addition to the Strong Nuclear force, through which the hadrons are produced (in future to be called Strong with a capital S) there is a completely separate disruptive Weak nuclear force lurking inside the hadrons, which eventually causes them to disintegrate into lighter particles. The strength constant (analogous to the fine structure constant for electromagnetic effects) has a numerical value of about one ten million millionth (10^{-13}), which is the factor by which the observed life times differ from the typical nuclear time.

Nobody can visualise one million and there are so many factors of a million involved in this argument that the reader is probably thoroughly confused. Let us summarise the findings so far in terms of the analogy with collisions of cars in a car park. We have discovered that at low energies for the stream of cars driven into the full park, the colliding cars bounce off each other with fairly violent action and reaction, but remain intact. As the speed of the stream of cars is increased, the colliding cars start to come to pieces and fragments such as doors and wheels are found as separate entities among the debris. This is the analogue of pion and *K*-meson production. At even higher energies a colliding Mini can come out of the collision as a Rolls-Royce. This process—highly improbable in real life—is our analogue of protons turning into other heavier baryons such as Λ or Σ particles. The typical time for such car collisions is a fraction of seconds, which is analogous to the nuclear τ_n.

Completely separate from these sudden Strong effects, associated with collisions, are the quite separate Weak phenomena. If the incoming stream is stopped, and the cars and debris of previous collisions are just left to stand quietly in the car park, in the course of time they will start to fall apart. Given enough time doors and wheels will simply drop off. But the forces at work here, the slow effects of corrosion, are quite different from the violent forces operative during collisions. The forces of corrosion are analogous to the Weak nuclear forces. The time scale for such disintegration effects is quite different from that appropriate to collisions. The remarkable thing about nuclear decay is how long it takes. If we keep our relative time scales correct the slow disintegration effects take thousands of years, compared with the seconds involved in collisions.

Not only have we found a completely new nuclear force, but there is also associated with it a new family of particles—the *leptons*—which are not affected by the Strong interactions, but appear through the Weak interactions as decay products. This is the role played by the muon, which was at first mistaken experimentally for the pion. In this way also the neutrinos appear and, as far as nuclear and sub-nuclear physics is concerned, this is the very subsidiary way in which electrons play a part.

The remaining particle on Table 2 is the photon which only takes part in electromagnetic effects, but in this way it interacts with all charged particles, whether they be hadrons or leptons. Electromagnetic interactions are intermediate in strength between the Strong and Weak nuclear interactions. (See Table 3.)

TABLE 3

The fundamental interactions of Nature, giving their intrinsic strength constants and the range over which they operate. An infinite range denotes a mutual force between particles which falls off as the inverse of the square of the distance between them.

Type	*Strength*	*Range metres*
Electromagnetic	$e^2/hc = 1/137$	∞
Gravity	10^{-38}	∞
Nuclear, Strong	1	10^{-15}
Nuclear, Weak	10^{-13}	10^{-15}

The experimental technique through which this information has been collected is illustrated in the plates. The basic tool is the proton (or electron) accelerator which has been described above. Once a bunch of protons in the machine has been accelerated to its top energy, it is steered by magnetic fields into a target of some material, such as hydrogen or beryllium inside the vacuum chamber. Large numbers of secondary particles—pions, *K*-mesons and anti-protons—are produced in the collisions which the protons make in the internal target through the Strong interaction, and in the early days experiments consisted in the study of the products of these collisions using electronic particle detectors. Nowadays it is more usual for these secondary particles to be steered and focused down vacuum pipes into an experimental hall by a series of magnetic and electric fields. These particle beams are designed in the first place to select secondary particles of definite momentum and then to separate off those of a particular type. In this way beams of pions, *K*-mesons or anti-protons fan out in the experimental halls like railway tracks in a shunting yard. Plate III is a plan view of the 7 GeV proton accelerator, NIMROD, at the Rutherford Laboratory, showing the machine ring and various beam lines into the experimental halls. Plate IV shows part of the main machine ring and a beam line of extracted particles. To protect the experimenter from stray radiation the beams must be enclosed in concrete tunnels. This can be seen in Plate V, which is a view of the central area of the main experimental hall, where one beam is split into three. The real experiments take place when these beams are directed into targets, usually of hydrogen. The effects of the collision may be studied using electronic detectors (Plate VI(a)), but the most vivid pictures are obtained when the target protons are those of the superheated liquid of a hydrogen bubble chamber. (See Plate VI(b).)

An example of a sub-nuclear collision in a hydrogen bubble chamber is shown in Plate VIII, in which the beam is of 16 GeV negatively charged pions. Some of these have gone through the chamber without incident, but one pion has made a direct hit on a proton and produced the spectacular burst of the recoiling proton accompanied by eight positive and nine negative pions:

$$\pi^- + p^+ \longrightarrow p^+ + 8\pi^+ + 9\pi^-.$$

The event takes place in a strong magnetic field, which bends the tracks in circular arcs and, thereby, gives direct information on the particle momenta. This effect can best be seen by holding the picture just below eye level and looking back along the tracks towards the collision point. The direction of curvature determines the sign of the electric charge.

Most of the pions pass out of the picture without further interaction, but one has made three successive bounces off other protons in the liquid. The sudden change of direction of the pion track and the three heavy tracks of the recoiling protons are clearly visible. The small whirlpools are light spiral tracks made by stray electrons which happen to get knocked on by the pions.

A further example of this technique is reproduced in Plate IX(a). The picture shows the effects of bombardment of a proton in the liquid hydrogen by an anti-proton. On collision the proton-anti-proton pair annihilates and the energy is carried off by mesons which undergo a variety of subsequent Strong interactions or Weak decays. These are indicated in detail in the key shown on Plate IX(b). In symbolic form, using the particle labels of Table 2, the full story is the following:

$$\bar{p}^- + (p^+) \longrightarrow K^\circ \quad + \quad K^- \quad + \quad \pi^+$$

$$\pi^+ + (p^+) \longrightarrow p^+ + \pi^+$$

$$K^- + (p^+) \longrightarrow \Lambda^\circ + \pi^\circ$$

$$\hookrightarrow p^+ + \pi^-$$

$$\hookrightarrow \pi^+ + \pi^-$$

$$\hookrightarrow \mu^+ + \nu_\mu^\circ$$

$$\hookrightarrow e^+ + \nu_e^\circ + \nu_\mu^\circ.$$

Strong interactions have been indicated by bold arrows and Weak decays by a light arrow. Only moving charged particles leave visible tracks, which leads to a very clear distinction between Strong and Weak events. The former are collisions between a moving particle and a stationary proton in the bubble chamber. The latter does not give rise to a track, thus one actually sees only the tracks of the incoming particle and those charged particles which come out of the collision. The net charge on the particles is unchanged by the collision, so an even number of charged particles is involved. Since one of these does not give rise to a track, Strong effects always give rise to a vertex in the photograph at which an *odd* number of tracks converge. (We have indicated this by putting the struck target protons in brackets in the key above.) On the other hand the stationary protons in the liquid of the chamber are not directly involved in the Weak decays. All charged particles in a decay give rise to tracks. Charge conservation then requires that an *even* number of tracks must be seen to join at any Weak interaction vertex. An analysis of the length of tracks between the Strong production point and a particular type of Weak decay mode determines the mean life of the decaying particle.

A further example is given in Plate X(a) of a collision arising from a negative *K*-meson beam, which is interpreted in Plate X(b). The chain of events is the following:

$$K^- + (p^+) \longrightarrow \Sigma^+ + K^\circ + K^+ + \pi^\circ + \pi^- + \pi^-$$

$$\pi^\circ \rightarrow \gamma + \gamma$$

$$\gamma \rightarrow e^+ + e^-$$

$$K^+ \rightarrow \mu^+ + \nu_\mu{}^\circ$$

$$K^\circ \rightarrow \pi^+ + \pi^-$$

$$\Sigma^+ \rightarrow \pi^+ + n^\circ$$

Here three of the particles produced in the initial Strong interaction process are seen to decay via the Weak interaction. The π° disintegrates into two photons (γ) via the electromagnetic interaction. None of these leave tracks, but one of the photons has made an electron-positron pair in the electric field of a hydrogen atom in the liquid. It is interesting to look back in the foreshortened view along the π^+ track. The kink near the production vertex, where the parent Σ^+ decayed is then clearly visible.

The existence of neutral particles which leave no tracks is always deduced from the assumption of energy and momentum conservation at the vertices where it is evident from the charged particles that some interaction has taken place. Information about the masses of both charged and neutral particles is obtained through the requirement that energy is conserved at each collision, or decay. By such considerations the information contained in Table 2 can be built up from a systematic study of very large numbers of photographs.

These, then, are the main features of the first layer of reality lying beneath the atomic level, the details of which were mainly uncovered during the first decade after the Second World War, although of course the knowledge that there must be strong nuclear forces goes back considerably earlier, and the first technical application of nuclear physics was made for all to see over Hiroshima in 1945.

It should be stressed that the physical conditions in the biosphere, in which we live and to which most of our experiments are confined, are quite atypical of matter in the Universe generally. At the temperatures and densities with which we are familiar, nuclei consist to a very good approximation of only protons and neutrons. All the hadrons shown in Table 2 are produced by cosmic rays striking the Earth's atmosphere, but quickly decay and certainly on Earth the greatest intensity of matter in these exotic forms is to be found, man-made, at the big proton accelerators. However, in the Cosmos as a whole, particularly in its very early development and in the final stages of very large stars, the hadrons other than nucleons play a crucial role. Also, since the nucleons are blobs of sub-nuclear goo and interact through the exchange of sub-nuclear goo, an understanding of the structure of the nucleons depends on an understanding of the physical properties of the goo.

Nuclear Physics and the Stars

The most direct impact which these discoveries have on our general appreciation of the human condition is in the field of astrophysics, particularly in connection with the Sun. We have already seen that the basic quantity in physics is energy, and it is also evident that the basic source of energy for virtually everything which happens on the Earth is the Sun. (The ebb and flow of the tides due to the gravitational attraction of the Moon is one notable exception.) The sources of energy available to animal life on the Earth can all be traced back to the Sun. Green plants absorb solar energy in the process of photosynthesis through which carbon dioxide from the atmosphere is decomposed to form carbohydrates which are absorbed in the body of the plant. This energy can be utilised by animals when the carbon compounds recombine with other elements via the processes of digestion, circulation and respiration. Men also recover this stored solar energy by burning plants in the form of wood, coal or oil. Wind and water power comes much more obviously from the Sun. It is clearly solar energy which evaporates water from the oceans to set up the required cycle of rainfall and rivers, to drive the mill wheels and hydro-electric plants. It is again the Sun's heat which creates temperature differences in the atmosphere which are responsible for the winds put to use in windmills.

All this is straightforward. The physical puzzle arises when one calculates the energy which the Sun has produced. This turns out to be enormous. It is clear that only a tiny fraction of the total energy emitted by the Sun actually falls on the Earth yet in one second this amounts to more than man has obtained from fossil fuels in his whole history. After allowing for absorption by the atmosphere, the solar energy falling every month, at the Equator, on a small garden (10 metres × 10 metres) is equivalent to the complete combustion of one ton of coal.

The next significant fact is that this prodigious release of energy has been going on for a very long time. The existence of plant and animal life on the Earth's surface depends very critically on the temperature and either doubling or halving the radiation would make life on Earth impossible. The fossil records show no indication of any change of this magnitude and it seems safe to assume that the Sun has been shining, more or less as it does now, for a period which must have started at least as long ago as the formation of the Earth's crust.

One of the best estimates of this is interesting in itself because it depends on the nuclear effects which we are considering. We have seen that the chemical properties of an element are determined by the number of electrons in the normal atom, which in turn is fixed by the number of protons in the nucleus. The number of neutrons which combine with these protons to form a stable nucleus is found experimentally to be roughly equal to the number of protons. It is always possible to form unstable nuclei of the same chemical element by adding or subtracting a few neutrons. The alternative nuclear forms of the same chemical element are called isotopes. Unlike the stable form these unstable isotopes disintegrate, usually in one of two standard ways. A neutron inside the unstable isotope can turn into a proton with emission of an electron and neutrino by the standard Weak interaction process. Alternatively, the electrical repulsions inside the nucleus can eject a charged nuclear fragment which almost invariably takes the form of an α-particle, or Helium nucleus. This consists of two protons and two neutrons which bind together exceptionally strongly under the action of the nuclear forces. (The α-particles in Rutherford's experiment came from just such a natural radioactive source.) The mean life for an unstable isotope depends very critically on the interplay between nuclear attractions and electrical repulsions within the nucleus, which vary in detail from case to case. Consequently mean lives can vary from a fraction of a second to millions of years. In the latter category are certain isotopes of carbon, uranium and thorium. If a quantity of uranium or thorium is lodged in the Earth's crust, it will be found subsequently in conjunction with its decay product. Since the decay involves the emission of charge, the decay products have nuclei of a different element, which can be identified and separated. From the relative proportions of the original substance and the decay product one can calculate accurately how long the decay has been going on. This has been compared aptly by Gammow to determining the age of a village (of known population and death-rate) from the weight of bones to be found in the cemetery. This technique of 'carbon dating' has proved very effective in archaeology. When applied to uranium or thorium in the Earth's crust the age of the Earth comes out to be at least a thousand million (10^9) years.

When one combines this lower limit on the Sun's life with the known rate of radiation, it can be calculated that since its formation the Sun must have given off an energy equivalent to about one thousandth of its total mass. We have seen that the energies involved in chemical reactions are completely negligible on the mass scale and this is, in fact, a million times greater than any energy release which could conceivably rise from any chemical process. It must be the Strong nuclear force which is operating to produce such an effect.

In recent years a beautiful and quite detailed theory of stellar evolution has been developed from a study of the interaction between the classical force of gravitation and the newly found nuclear force. The latter is intrinsically strong (that is the basic force between two nucleons is strong), but each nucleon only interacts with its near neighbours and the 'tennis ball' effect which stops one piling more than a certain number of nucleons into a given volume, prevents the force from building up above the intrinsic value. On the other hand the gravitational force, although it is intrinsically weak, is of long range. Each nucleon in a star interacts gravitationally to some extent with every other nucleon and the net effect in the centre of a body as big as the Sun (which contains 10^{56} nucleons) can be comparable in strength to nuclear effects. (This is not true of the long-range electrical forces, because the opposite charges on electrons and protons tend to cancel each other out.)

In broad and somewhat simplified outline it would appear that the galaxies are formed from a very thin cosmic gas of protons and electrons which gets drawn together into vast condensations under the gravitational attractions between the particles. The galaxies then condense still further under gravity into star clusters and finally into stars, of which the Sun is typical. These stars fall inwards under their own weight, the energy going mainly into heat at the centre, until a temperature is reached at which nuclear energy starts to be released. This produces an outward pressure to balance the inward pull of the gravity, and at the same time fuels the vast and continuous outpouring of energy from the star which is observed in the form of radiation.

The actual energy producing process is that of nuclear fusion in which the various isotopes of light elements, involving up to four nucleons, are formed from the protons of the stellar material. Imagine two enormously powerful magnets. When fairly far apart they exert little influence over each other since, to a very good approximation, the repulsion between their like poles will cancel the attraction between the unlike poles. However, if by chance they happen to come close so that opposite poles begin to match up, the attractive forces dominate the situation more and more strongly, and they finally rush together with a mighty clash of steel, giving off energy in the form of heat and sound. Subsequently the two magnets form a stable, tightly bound, system and it takes a considerable amount of energy to tear them apart. Nuclear fusion acts in just this way when the protons come together to form nuclei, except that the attractive nuclear forces, which come into play at very short distances, are far stronger than magnetic forces, so that the energy released (expressed as a fraction of the mass involved) is enormously larger than for the magnets. The specific chain of processes which leads to nuclear energy release in the Sun is the so-called proton cycle, which consists of the following processes.

$$\text{(i)}\ p + p \longrightarrow (pn) + e^{+} + \nu_{e},$$
$$\text{(ii)}\ (pn) + p \longrightarrow (ppn) + \gamma,$$
$$\text{(iii)}\ (ppn) + (ppn) \longrightarrow (ppnn) + p + p,$$

where the symbols have the significance as in Table 2.

p proton,	n neutron,
e positron,	ν neutrino,
γ photon,	

and two or more nucleons in brackets such as (pn) indicate that they are fused together under the action of the Strong nuclear force to form a tightly bound nucleus. This can be an isotope of either hydrogen or helium depending, respectively, on whether one or two protons are involved. Thus in the first stage (i), one proton out of a pair in close proximity converts via the Weak interaction into a neutron, a positron and a neutrino. The resulting neutron then fuses with the original proton to form the (*pn*) system, which is an isotope of hydrogen. The (*pn*) combination is lighter than the mass of an unbound proton and neutron by an amount equivalent to the energy necessary to tear it apart. This is almost exactly one thousandth part of the mass of the two protons we started from, which is just the magnitude of energy release we are looking for. In the energy balance between the initial and final states the energy lost in the fusing of the (*pn*) system appears as kinetic energy of the two final leptons (the positron and neutrino). Note that this process is a combination of Strong and Weak nuclear phenomena.

The second stage (ii) is somewhat simpler. The (*pn*) system fuses with another proton from the stellar material to form an even tighter bound nuclear system (*ppn*)—an isotope of helium —the further mass defect in the bound system being balanced by the emission of a very energetic photon. This process involves Strong nuclear and electromagnetic forces.

Finally in the third stage (iii) two (*ppn*) combinations, arising from stage (ii), fuse to form (*ppnn*)—the normal helium nucleus or ubiquitous α-particle. This is a very tightly bound system so there is a yet further release of energy which appears as kinetic energy of the two stray protons. These contribute further to the heat of the stellar centre and are available to take part in subsequent (i), (ii), (iii) energy releasing cycles.

The net effect of this cycle of processes is that protons (hydrogen nuclei) are converted into α-particles (helium nuclei); the surplus charge is carried off by the positrons, and the loss of mass is compensated mainly by the large kinetic energy of the emitted photons and leptons. The positrons annihilate with stellar electrons to form high energy photons so the radiation, amounting to the required vast energy release, is emitted ultimately by the star in the form of photons and neutrinos. Thus common stars, and the Sun in particular, are seen to be enormous nuclear fusion reactors held together by the weight of their own fuel. In this sense our industrial civilisation, and indeed the whole life process on the Earth, has been running from the start on nuclear power.

It is important to notice that each member of the initial pairs of particles in the process (i), (ii) and (iii) is positively charged. Thus the members of the pairs repel each other and at normal terrestial temperatures the reactions will never take place. However, if the temperature is raised to ten million degrees, as it is in the centre of the Sun, the particles are moving fast enough to overcome the electrical repulsion, and in a sufficient number of random collisions they get close enough for the nuclear forces to take over, and the fusion processes take place. The high temperature originates from the gravitational collapse, but once the fusion has started, it feeds on itself maintaining the high temperatures and supplying far more energy than has been fed in to get it started. This is just the familiar effect of a coke fire burning hotter than a coal fire, but which requires some burning coal to get it started.

So far the only large-scale, man-made release of energy from nuclear fusion has been in the form of hydrogen bomb explosions which have energy yields equivalent to many million tons of TNT. There are already sufficient H-bombs in existence to wipe the human race off the face of the planet. The operating temperature of a fusion system is too great for it to be contained by material walls. The gravitational container which operates in the stars requires a system of stellar dimensions. For these reasons it has not yet proved possible to produce a sustained fusion reactor on the Earth, but intensive research is in progress to produce a device which can hold itself together by its own electromagnetic forces. Given reasonable priority this problem could almost certainly be solved by the end of this century, and one can hope that the effort will be maintained, since, ironically enough, in view of the bomb situation, here lies the only obvious hope for the long-term survival of humanity. In controlled fusion reactors one could burn the hydrogen from sea-water to form helium with an energy release, weight for weight, a million times more effective than in the combustion of coal or oil. Only with this essentially unlimited supply of energy will it become technically possible to irrigate and fertilise vast areas of the Earth's surface which are now barren and inhospitable, and so accommodate at a reasonable living standard a sensibly controlled total human population.

Nuclear Fission and Reactors

The trigger which is employed to set off terrestial fusion explosions is a nuclear fission explosion, colloquially and quite misleadingly known as an atom bomb. Certain isotopes of the very heavy, naturally occurring elements thorium and uranium, when they decay, do not invariably emit an α-particle, but instead split into two roughly equal portions. This takes place with a release which is not as great but comparable with that in fusion. This is what is meant by nuclear fission. It was discovered just before the Second World War that the fission of U_{235} (the uranium isotope consisting of 235 nucleons—92 protons and 143 neutrons) can be induced by bombarding it with neutrons. The energy so released appears in the motion of the fission fragments and the important point is that these fragments, in addition to two medium-size nuclei, include on average two or three neutrons. Clearly this opens up the possibility of a so-called chain reaction, since the neutrons produced by one fission in a piece of uranium may induce further fissions, which in turn produce more neutrons and so on. This can lead to a cumulative explosion of fission, like the population explosion depicted in the family tree of a prolific tribe in which every person produces on average two or three children.

The technical problem to be solved in constructing fission bombs, or nuclear fission reactors, is to produce circumstances in which the family increases and does not die out, as it will if on average each parent has less than one off-spring. Under these circumstances the chain breaks. Clearly the size of the system is crucial to the continuity of the chain, because a neutron can escape rather easily from a small quantity of uranium without causing a secondary fission. In a larger quantity it has further to go to get out, and the probability of inducing another fission is increased. The critical size is just that for which the probable number of secondary fissions induced by each primary fission switches over from being less than one to being greater than one, giving rise to a family which will increase, rather than die out.

Less than 1% of natural uranium contains the easily fissionable isotope U_{235}, over 99% being U_{238} (with three extra neutrons in each nucleus). For fission bombs the trick is to separate U_{235} from the rest to obtain a critical size of about five kilogrammes, which will explode to give an energy yield equivalent to some 20 000 tons of TNT. The separation is technically difficult, because the chemical properties of the two isotopes are identical, and they must be separated by some mechanical means which can distinguish the very small difference in mass. There are two methods. The one so far employed is essentially a filtering technique, based on the fact that the rate at which a nucleus will diffuse through a membrane depends on its mass. Hence the large diffusion plants in the nuclear club nations. The other possibility is by centrifuge, which separates by means of the centrifugal force which is greater on the heavier isotope. Either process is both difficult and expensive and provides a fortuitously formidable technical and economic barrier to the continued spread of nuclear weapons.

The first controlled nuclear reactor was built in a squash court in Chicago University in 1942 under the direction of the Italian physicist Enrico Fermi (known to his colleagues as 'the Pope' because he was always right!). This reactor is basically typical of other civilian nuclear reactors which have been constructed since to drive electricity generating stations. The practical restrictions on size and weight are much less restrictive for a reactor than for a nuclear weapon, since the system does not have to be transported. It proved possible to use natural uranium, of which only the 1% of U_{235} was useful fuel. This was like lighting a fire with coal which is very heavily contaminated with rock. The neutrons most effective in producing fission are slow, while those emitted in the uranium fission products are fast. To increase their chance of producing a fission these fast neutrons have to bounce around giving up energy to some substance such as very pure carbon (graphite) or heavy water, which does not absorb neutrons. This is called the moderator. In the original Chicago reactor the fuel rods of natural uranium were formed into an open lattice structure, buried in a huge pile of graphite. The critical size more or less filled the squash court and weighed about 1000 tons. The only other essential requirement is some control mechanism to keep the system operating at a constant rate, rather than growing by the family-tree effect into an explosion. This is done by means of control rods of cadmium which are strong neutron absorbents. When inserted into the pile, they suck up the neutrons and thus quench the fission-producing cycle. By a servo-mechanism these control rods can be continuously inserted or withdrawn in such a way that the neutron population is maintained at an essentially constant level, leading to a steady release of energy by the fission process. Once the critical size is reached, the system starts itself as a result of spontaneous fission. There is no problem of triggering as there is for a fusion reaction.

In the last twenty years there have been technical refinements in reactor design, but no changes in the basic ideas. There are some important cross-links between civil nuclear power and nuclear weapons programmes, because the fission reaction in the fuel rods of a reactor produces certain other fissionable elements, typically plutonium, which are not found occurring naturally. Since these differ chemically from other elements they are relatively easy to extract and so can form the basis of a nuclear weapons programme which avoids the extreme expense and difficulty of the U_{235}–U_{238} separation. But this takes us into the political, economic and military problems of nuclear technology, which is very far from our purpose and will not be pursued further.

The Role of the Weak Interaction

The study of Strong nuclear forces was undertaken to try to understand the existence of atomic nuclei. In this brief discussion of the role of nuclear phenomena in astrophysics and modern technology, it is clear that the Strong nuclear interaction does not only play a passive role in making atomic nuclei possible, it also plays an essential, dominating, active role in the stars, and technically it has the potential to transform completely human conditions on Earth—for better or for worse!

The Weak interaction is more mysterious because it was not anticipated, and it is not immediately obvious that it fulfils any essential function. If one imagines an all-powerful Creator who sets out to build the Universe from scratch more or less as we see it, he would clearly need protons, neutrons, electrons, pions and photons interacting through gravitational, electromagnetic and Strong nuclear forces. But would there be some point at which he would realise that the whole thing would not work unless the Weak interactions and the leptons were also included? This question is somewhat vague because it is hard to define precisely what is meant by 'more or less as we see it'. A more specific question is to ask, given the World as it is, what effect would it have on the proverbial man in the street if an all-powerful Creator were suddenly to switch off the Weak interaction? The result would, in fact, be drastic because the Sun and many other stars along with it would be disrupted. As we have seen, they now shine by the conversion of hydrogen into helium. This necessarily involves the conversion of protons into neutrons (see process (i) above), which can only take place via the Weak interaction. If this were stopped the continuous energy producing solar process would stop with it. Of course one reason that the stars are made of protons is that any neutrons which might occur in the primordial galactic gas would decay to protons, electrons and neutrinos by the Weak interaction before the galaxies had time to form. If the Weak interaction were switched off from the beginning of the Universe, this could not happen, and one could conceive of stars which were a mixture of protons and neutrons in the first place. In this case the fusion of nucleons into helium would go faster by a factor of at least a million since the Weak process (i) above could be replaced by the relatively rapid

$$p + n \longrightarrow (pn) + \gamma$$

which is similar to process (ii).

Another fascinating Weak interaction effect is the URCA process, which is a kind of thermostat which operates during the final stages of the life of a large star. A main sequence star like the Sun has a central temperature of about 10^7 K. When all the nuclear fuel is burnt, a large star starts to fall inwards under the gravitational attraction. The gravitational energy thus lost turns into heat and is trapped in the centre of the star, causing an enormous increase in temperature. When this gets up to 10^9 K the electrons are sufficiently energetic to induce the reaction

$$e^- + p^+ \longrightarrow n^\circ + \nu^\circ.$$

The neutron, once formed, decays via the standard Weak interaction: β-decay process

$$n^\circ \longrightarrow e^- + p^+ + \nu^\circ.$$

This gives back the electron and proton from which the process starts. Since the two neutrinos which have been formed on the way only have Weak interactions, they easily penetrate the material of the star and can escape. They thus carry off energy from the centre and prevent further heating. Otherwise the star is left as it was. Once the temperature drops below 10^9 K the process switches off. The general principle of the effect is just like an overflow pipe in a bath which only operates when the water reaches a certain level and then prevents it from getting any higher.

The name URCA refers to the Casino de Urca, which used to flourish in Rio de Janiero. It had a reputation for putting its customers through analagous cycle of operations which left them superficially unchanged, but drained all the surplus money out of their pockets!

Another effect of switching off Weak interactions, which is less drastic, but nearer home, is that the pion would become a stable particle. The stream of cosmic rays which strikes the Earth's atmosphere mainly in the form of high energy protons gives rise to pions, most of which now decay by the Weak interactions to muons. These muons undergo no further strong interactions and constitute 80% of the high energy ionising radiation which arrives at sea level. The most important function which these serve is to stir the gene pool of animal life, inducing mutations which make the process of natural selection possible. If the Weak interaction were switched off the strongly interacting pions would work their way down through the atmosphere losing most of their energy by constant sharing with other particles. This would probably be much less effective in producing mutations and, at the same time, all lepton emission from radioactive nuclei in the Earth's crust would also switch off. Thus without Weak interactions stellar evolution would be speeded up, while the whole development of life forms by natural selection would be slowed down.

It is evident from these considerations that, although the Weak nuclear interaction does not appear to be as vital to the scene as the other interactions, from either the biological or astronomical point of view, it does appear to determine the time scale of both stellar and biological evolution.

4

The Three Charges

Hadron Dynamics

Let us return to our main purpose which is to try to understand the nuclear interactions, which we have so far analysed only in a quite general way. This analysis has provided a qualitative distinction between the Strong nuclear forces which bind atomic nuclei together and produce the hadrons in nuclear collisions, and the Weak forces which cause the particles to decay. We must now consider these in much greater detail. Ideally we would like to find something equivalent in this context to the works of Newton and Maxwell on gravitational and electromagnetic forces. These are detailed mathematical statements of the way the forces operate, which enable one to predict with precision what will happen to any physical system dominated by these forces, provided it is sufficiently simple for the calculations to be tractable in practice. Eclipses can be predicted fairly simply on this basis. The use of computers has greatly extended the complexity of problems involving only gravity which can be subjected to such detailed analysis, and the precise handling of spacecraft on Moon flights, already referred to, is a dramatic example of this. A very important factor in such a situation is the number of bodies involved. If there are only two, moving under their mutual gravitational attraction, as in the case of the Sun and Earth, the problem can be solved exactly and is a standard part of any course in classical mechanics. If there is a third body present—if, for example, one wants the corrections to the Earth's orbit due to the presence of the Moon—then the problem must be tackled using approximate methods. Moon rocketry is clearly predominantly a three-body problem—the Earth, the Moon and the spacecraft—with the Sun producing minor perturbations. The standard technique in such cases is to start from the exact solution which most nearly resembles the physical situation. In the case of the Earth's orbit round the Sun, we can begin by neglecting the Moon altogether. This is already quite a good approximation since the mass of the Moon is only about one-eightieth of that of the Earth. One then calculates the additional effects which depend directly on this ratio of the masses, giving corrections of about 1%. The next corrections depend on the square of this mass ratio, which is less than one part in a thousand, and are therefore already quite small effects. In this way one gets the approximate answer as a series of terms, which are decreasing rapidly in magnitude. Although this is never exact, with enough work (which nowadays means access to enough fast computers) one can go to any desired accuracy.

In discussing the interaction of electrons and photons at relativisitic energies the forces are precisely defined by Maxwell's equations, but one never has the simplicity of a two-body system. This is because the interaction always allows the possibility of the emission of photons or the creation of electron-positron pairs, even if one starts from a simple situation in which, say, only two electrons are present. Consequently no problem in this field is exactly soluble. However, the strength of the interaction is determined by the fine structure constant e^2/hc, which is a small number (about 1/137). One can start from the trivial problem of two freely moving non-interacting electrons or the non-relativisitic problem of an electron bound to a proton, and then obtain a series of approximate solutions involving successively higher powers of the fine structure constant. This number is even smaller than the Earth to Moon mass ratio in the previous example so that, although the calculations are complicated, very accurate answers can be obtained from the first few approximations. This is the basis of very detailed theoretical results in this field, which are in very precise agreement with experiment.

In nuclear interactions one has none of these advantages. The simplest experiment with a high energy proton accelerator is to allow the proton beam to impinge on a hydrogen target, which can give rise to the whole variety of hadrons given in Table 2. This is already a many-body situation. The production of pions, *K*-mesons and anti-protons from the big machines is so intense that these particles from the primary collision can be channelled into secondary beams by a suitable arrangement of magnets and electric fields, as described in Chapter 3. These can be directed into hydrogen targets. In this way one can make a systematic study, for example, of pion-proton or *K*-meson-proton collisions, each of which can give rise to final states with many other particles present including the possibility of particle anti-particle pair creation. One is thus confronted with many, inter-dependent, many-body problems, all relating to the Strong nuclear interaction. The basic question is to find the general form of this interaction. But if one makes a guess at it, the only known way to calculate the implications of the guess is as an approximation in successive powers of the strength constant. Here one is foiled again, because one of the few things we do know is that this strength constant is of order one. This invalidates the entire method because one times one is again one. The second term of such a series of successive approximations, which should be a correction to the leading term, is as big as the first and so on. There is clearly no hope of arriving at answers in this way. With present mathematical techniques, we have no idea how to cope with this problem.

You might think that the Weak interactions are more tractable because they have a coupling strength which is small, and should provide a basis for an approximate solution. The difficulty here is that nearly all the known Weak decay processes involve at least two hadrons. (See Table 2.) These interact strongly with each other, and thus involve a combination of the Weak and intractable Strong interactions which confuses the situation. Notable exceptions are the decay of pions into leptons and the decay of a muon into an electron and neutrinos. This latter is the only process experimentally accessible at the moment which does not involve any hadrons at all.

Conservation of Energy and Neutrinos

In the last thirty years many abortive attempts have been made to develop alternative approximate methods appropriate to the Strong interaction many body problems. None can really claim to have had much success. What has been most successful is a more modest programme in which no attempt is made to determine the detailed dynamics of the nuclear interactions, but rather to consider their conservation laws. As explained earlier, this means a study of those properties of any nuclear system which remain unchanged throughout any interaction. These conserved quantities are, in turn, related to the transformations under which the interaction remains unchanged.

The most important of these conservation laws are the conservation of energy and momentum, and the very existence of neutrinos was deduced from an application of these general principles. The first Weak interaction process to be discovered was the decay of a neutron into a proton, electron and neutrino. This was seen to take place for a neutron bound in a nucleon. What was actually observed was that a nucleus, A, made a transition to another nucleus, B, with the emission of an electron. Let us assume that no other particles are involved. In the frame of reference in which A is initially at rest, the nucleus B must recoil in a direction immediately opposite to the electron, just as a boat recoils when you dive off the end of it. This follows from the conservation of momentum. By the conservation of energy, the kinetic energy of the electron and recoiling nucleus B, together with their rest energy, must be equal to the rest energy of A, which is the total energy to start with. This completely fixes the magnitude of the recoil. On this hypothesis all electrons should be emitted with the same energy. This is not what is seen. The electrons come off with a continuous range of energies, varying from zero to a maximum value equal to the one to be expected by the above argument, and in a direction not necessarily in line with the recoiling nucleus. At this stage physicists might have concluded that energy and momentum are not conserved. They followed the alternative explanation that another particle was being emitted along with the electron which was taking off the momentum and energy necessary to redress the balance and lead to over-all conservation. The particle must be electrically neutral to explain why it is not detected directly. If it were massive it would have to contribute at least its own rest energy to the final state. This reduces the maximum possible kinetic energy which can go to the electron. Since this value is not reduced one has to suppose that the extra particle has zero mass (like the photon). If angular momentum is to be conserved it must have a spin of $\frac{1}{2}h$ (like the electron).

The existence of the neutrino with these properties was postulated in 1933 by Pauli. It was extremely difficult to obtain direct confirmation of this hypothesis because the neutrino is the only particle which interacts with the rest of the world exclusively through the Weak interaction. All other particles have either electromagnetic or Strong interactions, or both, and consequently always interact in some way with quite manageable quantities of matter. They can be brought to rest even from high energies by a few feet of concrete. Neutrinos, on the other hand, can pass through dense matter at the speed of light for a whole year without any interaction taking place which would even show that they are there. However, according to the Pauli theory, a nuclear reactor should give off a vast flux of electrons and neutrinos, coming from the decay of all the excited nuclei in the fission products. Taking advantage of this, the neutrino induced reaction in which a neutrino and a proton combine to give a neutron and a positron,

$$\nu_e + p^+ \longrightarrow n + e^+,$$

was finally observed in 1953.

In the meanwhile Pauli's argument had become much more convincing by the discovery of such processes as the disintegration of a pion into a muon and a neutrino.

$$\pi^+ \longrightarrow \mu^+ + \nu_\mu.$$

Here again only the pion and muon are observed. This leads to a completely obvious and crazy breakdown of energy and momentum conservation, if it is assumed that they are the only particles present. The situation is perfectly straightforward if the neutrino is included.

Still later (1962) it was established that the neutrinos which are produced in association with muons, will subsequently induce the reaction

$$\nu_{\mu} + n \longrightarrow p + \mu^{-},$$

but cannot induce the similar reaction in which an electron is produced;

$$\nu_{\mu} + n \longrightarrow p + e^{-}.$$

From this one deduces that there are two types of neutrino, one connected with electron interactions and one with muons, as shown in Table 2. The decay process

$$\mu^{-} \longrightarrow e^{-} + \bar{\nu}_{e} + \nu_{\mu}$$

involves both electrons and muons and consequently both types of neutrinos.

Order, Disorder and the Stability of the Proton

Let us now consider in more detail the observed mean lives of the particles. We have said that the typical Strong nuclear time is

$$\tau_{n} \simeq 10^{-23} \text{ seconds},$$

whereas the typical mean lives are 10^{-10} to 10^{-8} seconds. From this we have deduced the existence of the Weak interaction with a coupling strength about 10^{-13} (the ratio of typical mean life to typical nuclear time). But there are some notable exceptions, namely the Σ°, π° and η° which decay much faster. All of these have photons among their decay products. Since the photon only operates through electromagnetic interaction, these necessarily bring in the fine structure constant $e^{2}/hc \simeq 1/137$. In the formula for the decay rate one such factor must appear for each photon. If the mechanism for the decay is purely electromagnetic, it would give these particles a fairly long life in terms of nuclear time, but kill them off faster than the much weaker Weak interaction, which is what is observed. This, then, seems to be the explanation.

The other odd-man-out is the neutron with a mean life of over ten minutes. This is a fantastically long time on the nuclear time scale and would appear to suggest a force even weaker than Weak. However, the answer is not so drastic, and we meet here in a simple form an effect which is very important in the subsequent discussion. The rate of decay of any particle depends partly on the strength of the interaction and partly on the 'amount of room' it has into which it can decay. Let us again go to cars for an analogy. If a car is parked out in the open, the speed with which it can make a getaway depends only on the power of the engine. However, if it is in a crowded car park the time may be appreciably extended, and if it is very tightly parked this may be the overriding consideration. A measure of the tightness of the parking is the number of alternative routes by which the car can be extricated. Similarly when a particle decays, the rate depends primarily on the strength of the interaction causing the decay, but is also influenced by the number of different configurations available to it in the final state, subject to energy and momentum conservation. These include the different possible orientations in space of the decay products, but the number also depends critically on the difference between the rest energies in the initial and final states. This in turn fixes, by the conservation of energy, the amount of kinetic energy in the final state. If the available energy is small it can be the dominating factor in determining the decay rate, but once it gets so large that plenty of routes are available—as in the case of the parked car—increasing the freedom still further does not make any difference. The neutron β-decay is an extreme case of tight parking. It has a rest energy of 0·9395 GeV, while that of the proton is 0·9382 GeV and the electron 0·0005 GeV giving a difference of only 0·0008 GeV. When allowance is made for this, it is found that the strength of the decay interaction turns out to be typically Weak.

Note that this delicate balance is very important to the details of nuclear stability. We have discussed the behaviour of a free neutron. When a neutron is bound in a nucleus it becomes effectively lighter, due to the binding energy, and for stable isotopes this β-decay mechanism is no longer energetically possible. Inside a stable nucleus a neutron remains indefinitely a neutron. However, many semi-stable isotopes decay by this means with mean lives varying from seconds to years depending on how little energy there is available—the tightness of the parking.

If the decay of a hadron involves leptons it is clear that the Weak interaction is involved in an essential way. But most of the remaining disintegrations listed in Table 2 involve hadrons decaying into hadrons, and there are two major questions which have to be answered. First, why is the proton stable and second, since in many cases only strongly interacting particles are involved, what is it which prevents the hadrons disintegrating via the Strong interaction?

Let us consider the stability of the proton first, since this is clearly crucial to the world as we know it. From the atomic point of view, the proton is one of the basic building blocks. Yet from the behaviour of the other hadrons on Table 2 there is no obvious reason why it should not disintegrate into, say, a positive pion and neutrino, which is not forbidden by any conservation law which we have so far introduced. How certain can we be that such a process is not possible?

To answer this we again require the notion of the number of states in which a system can exist. Consider an orderly living-room furnished with chairs, tables, cushions, books, ornaments, carpets and so forth. Each object in the room has a fairly well-defined place, and there are consequently a relatively limited number of states in which the room can be considered tidy and in good order. For every object in the room there are, of course, vastly many more positions in which it would be considered out of place. When these possibilities for all the objects in the room are multiplied together the number of untidy or disordered states exceeds the ordered ones by some enormous factor. If some mixing or shuffling element is let loose in the room, like a small child who is insensitive to the fine distinctions between order and disorder, it is absolutely certain that within a short time the room will be in a state of chaos. There is nothing magical about this. It follows inexorably from the sheer weight of numbers. The child operates without prejudice so that under his hand all states of the objects in the room are equally likely. Since the untidy ones predominate over the tidy ones by such a vast factor, it is absolutely certain that some untidy state will emerge. This purely statistical statement forms the substance of the Second Law of Thermodynamics* and every housewife spends her life fighting it. Since the First Law is just the conservation of energy, already incorporated in our basic scheme, this is the most important new principle, which emerges when one starts dealing with many body systems. This supplements, by statistical arguments, the primary physical laws so far considered, which refer to the detailed motion of each component of a system. The rate at which the numbers build up in the Second Law situation can be illustrated by considering a pack of playing cards. We can define an ordered, or tidy, state to be one in which the cards are arranged by value in successive suits. There are just twenty-four such configurations which arise from the different possible orderings of the suits. This is itself a surprisingly large number, but the number of different ways the fifty-two cards can be arranged is about a ten thousand million million million million million million million million million million million million million million (10^{52}). The chance of finding a shuffled pack in an ordered state is the ratio of these two numbers. Although, in principle, there is always some finite probability, in practice it is absolutely safe to neglect it and one can confidently assert that an ordered shuffled pack will never be found. It is to be remarked that no value judgement is

* The logarithm of the number of different states in which a system can be found is called the *Entropy*. Thus the entropy of tidy or ordered states is very much less than that of untidy or disordered ones.

involved in labelling the condition of a system ordered or disordered. By definition an ordered condition is one which can exist in relatively few configurations, whereas disordered is one which has many more—normally vastly many more—possible configurations.

The relevance of this to our problem is that one may think of a proton at rest as a very highly ordered condition of a certain amount of energy—the rest energy of the proton—which can exist in just one state (strictly two if we allow for two possible orientations of the proton spin). If the proton can decay by any mechanism into two or more lighter particles, these serve to define an alternative condition of the system which is relatively highly disordered, since it can exist with all conceivable orientations. The number of allowed states depends on the relative momentum of the decay products much as the number of points on the circumference of a circle depends on its radius. The decay interaction is the shuffling agent, corresponding to the child let loose in the drawing-room. If it exists and operates on a time scale comparable with the age of the Universe, then by the relentless operation of the Second Law, essentially every proton would by now have decayed into lighter particles, just as every living-room would surely be reduced to a shambles. Clearly the opposite is the case, and there must be some very exact law which is preventing this from happening.

Before pursuing this further let us just remark on the generality of the argument. Rest energy (or mass) is a very highly ordered condition. Disorder in a system—that is the number of accessible states—is always increased by any interaction which reduces the amount of mass present and breaks it up into a greater number of lighter particles, thereby turning a greater proportion of the total energy into kinetic energy. In terms of our analogy, this is just the very obvious remark that the child's power to produce chaos is enormously enhanced by an ability to reduce objects to fragments—to tear up books and smash ornaments—and the greatest disorder is produced when the smashing is most complete. Thus, as time goes by, any system will tend to degenerate into a condition with the minimum amount of mass, the largest number of parts and the maximum amount of motion allowed by the conservation laws. The historical development of the Universe is dominated by this statistical rule, operating within the framework of the primary physical laws which we are trying to determine. We come back to this in Chapter 7.

Put this way, it is clear that there is a general tendency for rest energy to convert to photons and neutrinos (the stars are doing just this), and from the point of view of its stability, the electron presents the same problem as the proton. An electron is coupled by the Weak interaction to neutrinos, and one must ask why electrons have not by now all converted into neutrinos and kinetic energy. The reason for this is clear and leads us to an answer to the other two problems.

Electric and Baryon Charges

In any electrical circuit it is found that the net total current into a junction is zero. Since the current is just the flow of electric charge, this implies that the net flow of charge into any junction must be zero. In Maxwell's equations this rule is generalised to the statement that the change in charge in any small volume is just balanced by the net flow of charge into the region. More simply one can say that charge is never created or destroyed. If one thinks of this in microscopic terms where the charge is always carried by the particles in integer multiples of the charge on the electron, it means that in any electromagnetic process the total charge of a closed system remains unchanged. Electric charge, like energy and momentum, is conserved.

It is observed that this law remains valid when we include the effects of the other interactions—gravitational, nuclear Strong and Weak. From the point of view of charge conservation and elementary particles the physical properties of electric charge are amazingly simple. We do not have to bother with electromagnetic effects such as the attraction between unlike charges or the emission of photons by accelerating charges. The electric charge is just a numerical label attached to each particle, taking some

This is a view of the centre of the main NIMROD experimental hall, shown in the plan of Plate III, where one beam line splits into three. To protect the experimenters from radiation the beams are buried in concrete tunnels. One beam line has been exposed for modification in the left foreground of the picture.

(Photo: Rutherford Laboratory)

(a) A typical electronic counter assembly. A beam of pions of selected momentum enters the apparatus down the pipe from the right, where it strikes a hydrogen target. The cylindrical array of counters is designed to detect charged pions and photons from neutral pions which are produced in the pion-proton collisions in the hydrogen. *(Photo: Rutherford Laboratory)*

(b) A view of the British national hydrogen bubble chamber built during the sixties. The picture shows the nine illumination ports to the liquid hydrogen which has an operative volume of $150 \times 45 \times 45$ cm^3. Very much bigger chambers, operating in larger magnetic fields through the use of superconducting magnets, are under construction. *(Photo: Science Research Council, UK)* simple value such as zero, or plus or minus one, and the conservation law asserts that the sum of these labels remains unchanged throughout the development of any system. This imposes a large number of simple restrictions on the processes which can take place. Thus we can have, for example,

$$\pi^- + p^+ \longrightarrow n^\circ + \pi^\circ,$$

or

$$K^- + p^+ \longrightarrow \Lambda^\circ + \pi^\circ,$$

but the decay of a charged particle into only neutrals is impossible; for example,

$$\Xi^- \not\longrightarrow \Lambda^\circ + \pi^\circ.$$

The conservation of electric charge immediately frustrates the Second Law as far as the disintegration of the electron into neutrinos is concerned, because the electron is charged, whereas the neutrinos are neutral. The disintegration is quite consistent with energy and momentum conservation, and it would greatly increase the disorder, but it is flatly forbidden because the neutrinos cannot carry off the charge. The lightest particle carrying any such conserved charge must always be stable.

Note that this rule does not forbid the annihilation of an electron-positron pair into photons. Here the total electric charge is zero. This is, in fact, a classical example of statistics favouring the state with the least rest mass. There are many more configurations available to the system in its two photon form than in its electron-positron form which is why, given a chance, annihilation always takes place.

To ensure the stability of the proton we introduce another conservation law of exactly the same structure as the conservation of electric charge, in the form in which it applies for the nuclear interactions. To each particle we attribute another label specifying its 'baryonic charge' which, like electric charge, can be any integer, but usually has some simple value such as zero or unity. We attribute baryon charge of plus one to all the baryons, minus one to all anti-baryons and zero to all the other particles in Table 2. We postulate further that the net baryonic charge is exactly conserved for any physical system. This law is found to be satisfied. For example, we see the decay

$$\Sigma^- \longrightarrow n + \pi^-$$

in which the baryon charge of one is conserved. But the decay into two pions is not observed;

$$\Sigma^- \not\longrightarrow \pi^\circ + \pi^-.$$

This decay mode involves a much greater reduction in rest mass and would therefore be statistically favoured if it were possible. Since the proton is the lightest particle to carry non-zero baryon charge, it is made absolutely stable by this law, in accordance with observation.

This law may be made to appear less arbitrary in terms of the analogy of the collisions of cars in the car park. We have seen that when the collisions get sufficiently violent (Strong effects), or if the cars are simply left to rot for a sufficiently long time (Weak effects), they will start to come to pieces. All that the law of baryon charge conservation is saying is that, however violent the collision, or however long you wait, the total number of objects which can reasonably be regarded as cars (baryons) remains constant. Neither in a collision nor in a decay does one ever get the total disintegration of a car into fragments. For each car at the start there is always one object at the end which is recognised as what is left of the original car.

Hypercharge in Strong and Weak Interactions

It is found that the Strong collisions show yet further special features along these same lines. We have compared the production of Σ or Λ baryons in proton-proton collisions with the appearance of a Rolls-Royce from a crash between two Minis. Whenever this happens it is found that the Rolls is always accompanied by, say, a door among the debris and never with a wheel or a back axle. We can systemise these regularities by attributing to each hadron yet another charge, the 'hypercharge', Y; which also takes on some integer value, quite independent of the electric and baryon charges. The total hypercharge is also conserved in all Strong interactions. The values of the electric charge, Q, baryon charge, B, and hypercharge, Y, for the hadrons of Table 2 is given in Table 4, and in the processes discussed below we indicate the hypercharge by a suffix to the particle symbol, for example, π_0^- implies a pion with negative electric charge and zero hypercharge. The conservation of hypercharge ensures, for example, that in pion-proton collisions (a process which can be studied in the big accelerator laboratories) the production of Σ's and Λ's must always be accompanied by either positively charged or neutral K-mesons. One of the most striking effects of

TABLE 4

The electric Q, baryon B and hyper-Y charges of the least massive hadrons. The same pattern of Q and Y values appears for the baryons and for the mesons. This is displayed in the Q–Y plots of Fig. 3.

	Q	B	Y
Baryons, Spin $\frac{1}{2}h$			
Ξ	0	1	−1
	−1	1	−1
	+1	1	0
Σ	0	1	0
	−1	1	0
Λ	0	1	0
N (p)	+1	1	+1
N (n)	0	1	+1
Mesons, Spin zero			
$\bar{K}$	0	0	−1
	−1	0	−1
	+1	0	0
π	0	0	0
	−1	0	0
η	0	0	0
K	+1	0	+1
	0	0	+1

hypercharge conservation which does not follow from any other law is that the process

$$\pi_0^- + p_1^+ \longrightarrow \Sigma_0^- + K_1^+$$

is allowed, whereas the very similar process,

$$\pi_0^- + p_1^+ \not\longrightarrow \Sigma_0^+ + K_{-1}^-,$$

is forbidden (since in the latter process the hypercharge would have to switch from plus one to minus one). This is exactly in accordance with what is observed.

Hypercharge conservation supplies the answer to our question why the hadrons of Table 2 do not decay via the Strong interaction. All decay routes involving only hadrons in the final state which are allowed by energy conservation (a decaying particle must be heavier than its decay products), involves a change in the total hypercharge. This can easily be checked from Tables 2 and 4. For example, the decay

$$\Lambda_0^\circ \longrightarrow p_1^+ + \pi_0^\circ$$

involves a change of hypercharge from zero to one. This means such decays are forbidden for the Strong and electromagnetic interactions, but evidently can go through the Weak. The conservation of electric charge and baryon charge are absolute physical laws, necessary to ensure the stability of the electron and proton, and thus give us the world as we know it. In contrast the conservation of hypercharge is an approximate conservation law which is respected by the Strong and electromagnetic forces, but is violated by the Weak interactions.

One might expect that since the Weak interaction does not conserve hypercharge, it would not be concerned with it in any way. It is observed, however, that although the total hypercharge can change in a Weak interaction, it only changes by one unit. Thus

$$\Sigma_0^+ \longrightarrow p_1^+ + \pi_0^\circ$$

and

$$K_{-1}^- \longrightarrow \pi_0^- + \pi_0^\circ$$

both involve such a change. In these cases the operation of the rule is not very striking because the decay patterns are already determined by the necessity of conserving energy, electric charge and baryon charge. A more significant example is the decay of the Ξ^- baryon, which is seen to decay via the route

$$\Xi_{-1}{}^{-} \longrightarrow \Lambda_0{}^{\circ} + \pi_0{}^{-},$$

but never

$$\Xi_{-1}{}^{-} \not\longrightarrow n_{+1}{}^{\circ} + \pi_0{}^{-}.$$

This is remarkable since from the point of view of the number of configurations available (tight parking) the latter decay scheme is strongly favoured, because the neutron is appreciably lighter than the Λ. The only reason that this process does not take place is that it involves a change of hypercharge by two units.

The Strong and Weak nuclear interactions operating among the variety of particles listed in Table 2, subject to the restrictions imposed by the three types of charge, provide us with a fascinating world of new sub-nuclear phenomena. These are seen most vividly in photographs taken in hydrogen bubble chambers. A very beautiful example is reproduced in Plate IX. The protons in the chamber constitute a large tankful of target protons which sit there like the cars in our car park. The picture shows the effects of bombardment by a high energy anti-proton. On collision with a proton in the chamber it annihilates into mesons which undergo a variety of subsequent Strong interactions or Weak decays, which are indicated in detail in the key to the figure. Only moving, charged particles leave visible tracks, which provide a very clear distinction between Strong and Weak events. The former are collisions between some moving particle and a stationary proton in the bubble chamber, which does not give rise to a track. Thus one actually sees only the track of the incoming particle and those charged particles which come out of the collision including the recoil proton (or other baryon into which it has been converted).

Consequently Strong effects always give rise to a vertex in the photograph at which an *odd* number of tracks meet. On the other hand the protons in the chamber are not directly involved in the Weak decays. All charged particles involved in a decay give rise to tracks. Charge conservation then requires that an *even* number of tracks must be seen to join at any Weak interaction vertex. The interested reader may be amused to check this rule in the interpretation of Plate IX (and Plates X, XI and XII) and to confirm that whereas all charges are conserved at the Strong vertices, hypercharge is lost or gained by one unit at the Weak points.

The existence of neutral particles which leave no tracks is always deduced from the assumption of energy and momentum conservation at the vertices where it is evident from the charged particles that some interaction has taken place.

We shall come back to the subject of the three types of charge, which provide a clue to a much deeper understanding of the Strong interaction, but let us for the moment pursue further the study of the Weak interaction.

5

Through the Looking Glass

Space Reflection and Parity

We have already seen the dominant way in which the conservation of energy and momentum come into the analysis of sub-nuclear events. It was described in Chapter 1 how these laws follow from the very general assumption that the physical development of any closed system is invariant with respect to translations in time and in space. The very concept of universal physical laws, which should be valid for all places and at all times, already implies that the total energy and the total momentum of a closed system should remain unchanged.

We may profitably consider whether these primary laws are also invariant with respect to other possible changes. Suppose, for example, we imagine that instead of looking directly at our experimental apparatus, it were always viewed reflected in a mirror. One can ask whether under these circumstances one would deduce exactly the same primary physical laws. This is, of course, the situation made famous by Alice on her trip through the looking glass. But Lewis Carroll allowed himself considerable licence, and many things happen on the unphysical side of Alice's mirror, which are not logical consequences of living in a world in which left and right have been interchanged.

The required situation is easily visualised as that which one would see if an ordinary film were shown with the light from the projector coming through from the wrong side. Any peculiarities which appear would be a measure of the extent to which everyday life is not invariant with respect to space inversion. It is clear that for the most part things would look much as usual, but certain details would be changed. Clocks would go round the wrong way, the majority of screws would be seen to have left-hand threads, and most people would write left-handed and greet each other by shaking the left hand. This conventional interaction between human beings could be made invariant either by introducing a Roman style handshake in which each person uses both hands, or by making a left- or right-handed shake equally conventional. In the latter case one style would go into the other on reflection, but the total impression would be unchanged.

All classical laws concerning gravity and electromagnetism are unchanged by such a reflection, and exactly the same physical laws would have been found if all experiments had been analysed in this perverse manner of viewing in a looking glass. This means that there is no absolute difference between left and right. Having established by convention on Earth what we mean by a right-hand thread, it is impossible to give a physical definition to thinking beings on a distant galaxy which would ensure that, without reference to anything outside their galaxy, they also employed the same convention. This is the analogue in terms of reflection of the fairly obvious statement that position has no absolute significance, if a closed system is invariant with respect to displacements. It is impossible to determine the position of a closed system without reference to something outside it.

As is often the case, it is easiest to see the significance of invariance with respect to space inversion by imagining how it could be broken. The law of gravity states that masses attract each other, and in a system consisting of just two bodies they would be seen to draw together under their mutual gravitational attraction. In a mirror this would look the same. It is conceivable that, instead of just drawing together, the two bodies should come together with a screwing motion. In this case each body would rotate in a right-hand direction about the axis joining them, like two nuts being screwed towards each other on the shaft of a bolt with a right-hand thread. This modified gravitational law would not be invariant under reflection, since in a mirror the bodies would appear to approach each other with a left-hand rotation. With such a law the absolute definition of a right-hand thread would be completely trivial. For all places and all times it could be defined as the 'handedness' with which two bodies screw together under their mutual gravitational attraction. This example also makes it clear why such a law is contrary to the normal expectations of Newtonian theory. From Newton's point of view massive solids are a conglomeration of structureless point particles, and the gravitational attraction between the bodies is the resultant force made up of all the separate attractions between the constitutents. In order to generate the screw motion of the massive bodies, it is necessary to assume such a motion for the point particles, but for Newtonian bodies the rotation of a structureless point is absurd.

In quantum mechanics the invariance of physical laws with respect to space reflection is more fruitful than in classical theory. This is because a particle's history is not described by an orbit which fixes exactly its position and momentum at any instant, but by a state function whose square determines the probability distribution of these variables. If the laws which control the motion are invariant with respect to space inversion, this distribution is unchanged by reflection. This means that on reflection the state function must either remain exactly the same, or remain the same apart from an over-all change of sign. This property of the state function—whether or not it changes sign under space reflection—is known as the *parity*. As time goes by the state function changes, but if the forces operating are invariant with respect to reflection, the parity does not change. It is conserved throughout the motion of the system. Thus for quantum systems invariance with respect to space inversion leads to a conservation law, just like invariance with respect to displacements. The Strong nuclear interactions like gravity and electromagnetic effects are found to satisfy this invariance property, and the fact that parity is conserved is of considerable assistance in the analysis of Strong nuclear processes.

The conservation of parity is particularly simple when applied to the decay of a particle such as that of a pion into two photons, which takes place through the Strong and electromagnetic interactions. The two photons emerge in a state of negative parity—one that changes sign on reflection—so by the conservation law the pion itself must have negative parity. This is an intrinsic property of the pion, like mass and spin, which it carries around with it wherever it goes, and serves to distinguish it from other particles.

When similar considerations were applied during the early fifties to the decay of charged *K*-mesons into two or three pions, a strange effect was observed. On detailed analysis the states of the decay products with two pions were found to have the opposite parity to those with three pions. If parity were conserved in these Weak interactions, this would imply that there would have to be two types of *K*-meson of opposite intrinsic parity, one decaying into two pions and the other into three pions. This would have been most peculiar since all other properties of the particles—their mass, spin and mean lifetime—proved to be the same. It was finally suggested that possibly the Weak interaction was not invariant with respect to space reflection, and it was realised that parity conservation had become so much an article of faith among physicists that no direct check of its relevance to the Weak interaction had ever been made.

We have seen that non-invariance with respect to space reflection reveals itself explicitly through some 'handedness' which relates a rotation, or spin, to a direction along the axis of rotation. This determines either a left-hand or right-hand motion. Since the *K*-mesons and pions both have zero spin, we must look at some more complicated system for a direct observation of the breakdown of space reflection invariance. What was done was to study the angular distribution of electrons in the β-decay of an isotope of cobalt, Co^{60}. The cobalt nuclei have spin, and at low enough temperatures it is possible to align the spins by applying a strong magnetic field. It was observed that the electrons are emitted preferentially in the direction of the axis of nuclear spin, in a way which by convention we describe as that in which the nucleus spins *left-handed* (anti-clockwise) about the preferred direction. This property of the decay is in clear violation of invariance with respect to reflection, because by the same convention in a mirror the nucleus would be seen spinning *right-handed* about the preferred direction. (See Fig. 1.) In terms of the state function, this implies that parity is not conserved. Very shortly after this discovery similar effects were observed for the electrons in the decay of muons. The news of these experiments broke in the *New York Times* in January 1957.

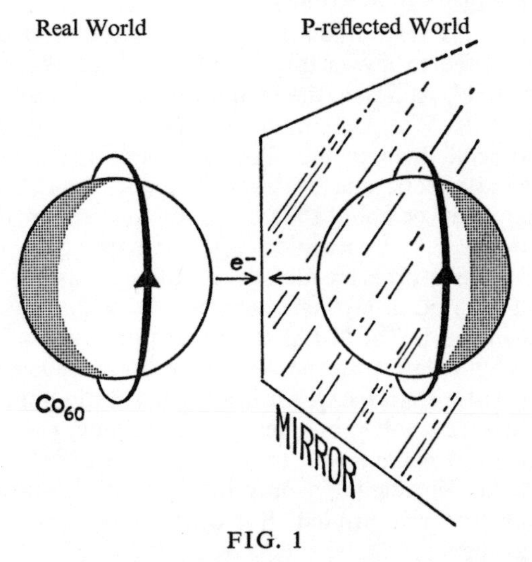

FIG. 1

In the real world electrons are emitted preferentially by Co_{60} in such a way that the nucleus spins left-handed about the preferred direction. In the reflection the spin appears right-handed about the preferred direction.

At this time the grand old man of high energy physics was Wolfgang Pauli, then at the Technische Hochschule in Zürich. He kept abreast of every development in the field, and it was standard practice to send him advance copies of papers for publication in the hope of catching his critical eye and provoking some favourable comment. It was a risky business since Pauli's adverse criticism could be very biting. Pauli, having originally proposed the neutrino to conserve energy, momentum and angular momentum in β-decay, was naturally fascinated by the puzzles arising over space reflection invariance. He had expressed very forcefully the opinion that this should be a property of all primary interactions. He firmly believed that parity must be conserved. His reaction to the news of the experiments in a letter to Professor Weisskopf at MIT, vividly expresses the surprise felt by most physicists of the time, and gives some feeling for the way such research is carried out. Pauli wrote:

> Now the first shock is over and I begin to put myself together again (as they say in Munich).
>
> Yes, it was very dramatic. On Monday 21st at 8.15 p.m. I was supposed to give a talk about 'past and recent history of the neutrino'. At 5 p.m. the mail brought me three experimental papers: C.S. Wu, Lederman and Telegdi; the latter was so kind to send them to me. The same morning I received two theoretical papers, one by Yang, Lee and Oehme, the second by Yang and Lee about the two-component spinor theory. The latter was essentially identical with the paper by Salam, which I received as a preprint already six to eight weeks ago and to which I referred in my last short letter to you. (Was this paper known in the U.S.A.?) (At the same time came a letter from Geneva by Villars with the *New York Times* article.)
>
> Now where shall I start? It is good that I did not make a bet. It would have resulted in a heavy loss of money (which I cannot afford); I did make a fool of myself, however (which I think I can afford)—incidentally, only in letters or orally and not in anything that was printed. But others now have the right to laugh at me.
>
> What shocks me is not the fact that 'God is just left-handed', but the fact that in spite of this He exhibits Himself as left/right symmetric when He expresses Himself strongly. In short, the real problem now is why the Strong interactions are left/right symmetric. How can the strength of an interaction produce or create symmetry groups, invariances or conservation laws? This question prompted me to my premature and wrong prognosis. . . . I don't know any good answer.

The experimental papers mentioned by Pauli are the ones we have described. The theoretical ideas to which he referred provided, as it turned out, a very elegant framework in which to interpret the results, exploiting to the full the peculiar characteristics of the neutrino, which enters physics only through the

Weak, parity violating, interaction. If we have a particle with spin, we can consider a state in which the sense of the spin is, say, left-handed about the direction of motion. But for a massive particle this is not a well-defined situation, because the particle must be travelling with a velocity less than that of light and the situation just described must be as seen by some particular observer. It is always possible to find another observer, who overtakes the particle, so reversing its relative direction. However, both observers agree about the direction of the spin, so the second observer will see the particle spinning right-handed about its direction of motion with respect to him. One may visualise the particle as a cyclist and the first observer as a pedestrian following him down a straight road. The second observer is the driver of a car which overtakes the cyclist. If the cyclist (to help us!) wears his watch on his forehead, it will appear with hands rotating left-handed about his direction of motion relative to the pedestrian, but right-handed about his motion relative to the car. (See Fig. 2.)

But a neutrino which has zero mass is quite different. In order that it should have non-zero energy and momentum it

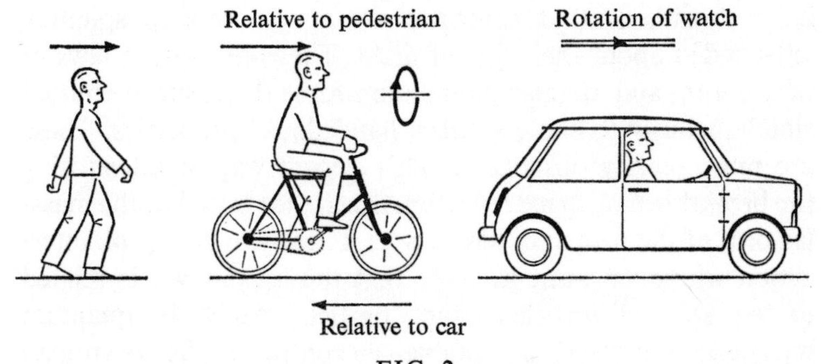

FIG. 2

The hands of the watch on the cyclist's forehead rotate left-handed about the direction of motion relative to the pedestrian, but right-handed about the direction relative to the car. If the cyclist goes with the speed of light, this ambiguity is eliminated and the 'handedness' is the same for all observers—and reverses when viewed in a mirror. This is the basis of the two component neutrino theory of the breakdown of reflection invariance.

must travel with the speed of light (the relativistic equivalent of infinite velocity). No observer can overtake a neutrino so that one which spins, say, left-handed about its direction of motion does so for all observers. It is thus perfectly consistent with relativity to postulate that only left-handed neutrinos exist in Nature, and that the Weak interactions are confined by this restriction. This is clearly not consistent with invariance with respect to reflections because in the reflected world one would find only right-handed neutrinos—a readily distinguishable situation. It is hard to observe the left-handedness of the neutrinos directly, because they are so elusive. But the fact that the Weak interaction involves only such neutrinos fixes its form almost uniquely. It then leads to observable effects which are not invariant for reflection and in particular makes specific predictions for the decay of Co^{60} and the purely leptonic process $\mu^{+} \longrightarrow e^{+} + \nu_{e}^{\circ} + \bar{\nu}_{\mu}^{\circ}$ in detailed agreement with experiment.

If one thinks of the neutron or the muon in these processes as a neutrino gun, the reflection invariance principle would require that it should be a cannon with no rifling in the barrel. What we have found is that, in fact, the barrel is always rifled with a left-hand thread, so that neutrinos are always emitted spinning left-handed about their line of flight. The conservation laws of momentum and angular momentum force the electron emitted simultaneously to show similar 'handedness' properties. These are more readily observable. This elegant way of introducing the breakdown of space reflection invariance is tied to the masslessness of the neutrino, which puts it neatly in the Weak interaction where we want it. Note that the breakdown is related to the spin of an elementary particle, which in quantum mechanics is perfectly acceptable—in contrast to the Newtonian situation considered earlier.

Charge Reflection

Although the naïvely interpreted space-reflection principle has broken down for Weak interactions, there is a more subtle reflection principle which could be satisfied. In the parity non-conserving neutrino theory, just described, there are only left-handed neutrinos. The right-handed neutrinos are excluded as physical particles. However, the theory allows anti-neutrinos, which must be right-handed and never left-handed. If we have a physical system in the real world which is made up of particles, it is purely a matter of convention whether we interpret its reflection in a mirror as a system of particles or the corresponding anti-particles. One is so used to the normal convention that the alternative interpretation is somewhat startling, but it is clearly a possibility. The only way to check would be to bring the real system and the reflection into physical contact and see whether they annihilate each other. Clearly this cannot be done except in science-fiction. The plot of some latter-day *Alice through the Looking Glass* using the anti-matter interpretation of reflections would be some macabre tale about a type of post Third World War situation, following the stupendous explosion caused by the almost total conversion of about a hundred kilogrammes of matter and anti-matter—Alice and anti-Alice—into heat, blast and radiation!

The usual reflection, which only involves a change from right-handedness to left-handedness, is called a P-reflection—P referring to parity. The alternative matter-into-anti-matter interpretation of what one sees in a mirror implies a reversal of the sign of the electric charges of all particles and is usually called a CP-reflection—C standing for charge. Let us consider what happens to our symmetry-breaking neutrino theory under CP-reflection. In the CP-mirror the reflection of a left-handed neutrino is interpreted as a right-handed anti-neutrino. By looking always in the CP-mirror one would come out with the primary law that there are only right-handed anti-neutrinos and left-handed neutrinos. This is the same as the law we started from. Thus this form of the Weak interaction is invariant for CP-reflection. This is analogous to the everyday situation in which right-handed and left-handed greetings are equally acceptable, the one going into the other on reflection to give the same over-all situation. The cobalt β-decay experiment tells us that the nucleus spins left-handed about the preferred direction of electron omission. In the CP-mirror we see what we interpret as anti-cobalt nuclei spinning right-handed about the preferred direction of emission of positrons. This is just what is predicted by the left-handed neutrino theory, showing again that the interaction is invariant for CP-reflection.

The invariance of all primary interactions with respect to CP-reflections again makes it impossible to give to an independent observer in a closed system a physical definition of a left-hand thread unless there is some independent way of deciding what is matter and what is anti-matter. Of course, once this is done, the required definition is just the 'handedness' of the neutrino spin.

Breakdown of CP-Reflection Invariance

This idea of invariance with respect to CP-reflections appeared for a few years to be in very satisfactory agreement with observations on the rather strange decay properties of neutral *K*-mesons into two and three pion states. One can define CP-parity in a manner exactly analogous to ordinary parity for conventional P-reflections, and this is conserved if all interactions are invariant with respect to CP-reflections. States containing two pions have positive CP-parity. The quantum combination of K° and $\bar{K}^{\circ}$ which has positive CP-parity is called K_1 and decays with CP-conservation into two pions with a typical mean life of about 10^{-10} seconds. The other independent combination with negative CP-parity, called K_2, can only go into three pion states. The kinetic energy available in this final state is much less—for this decay mode the *K*-meson is much more 'tightly-parked'—so it takes considerably longer to disintegrate and has a mean life about one hundred times greater. This very special situation of two completely distinct mean life times being associated with a neutral *K*-meson is in excellent accord with observation. By the conservation of CP-parity the long-lived neutral K_2-meson should never be seen

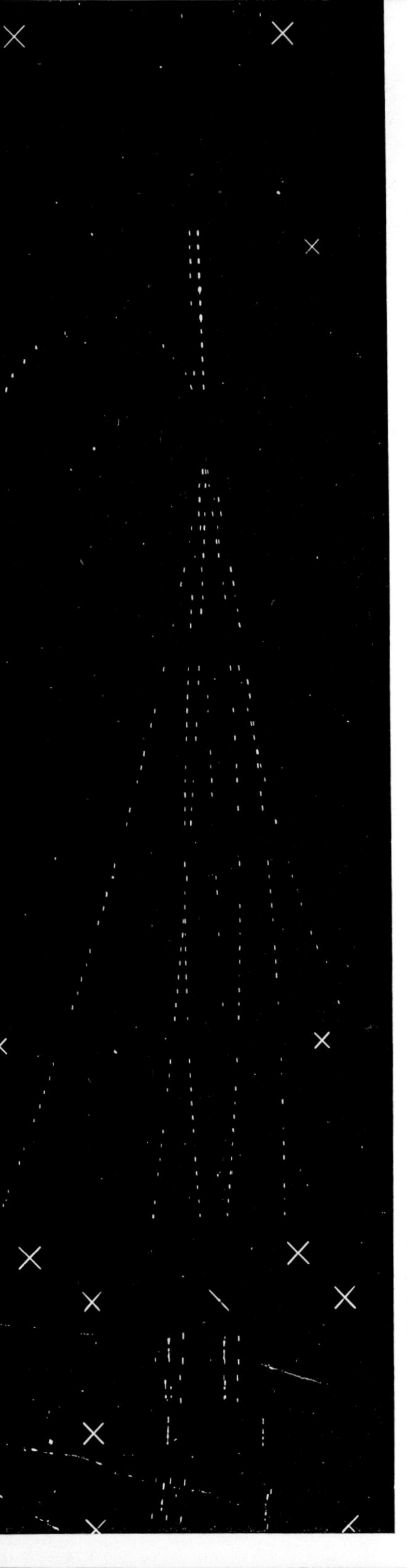

A spark chamber photograph of the annihilation of an anti-proton (single track from above) with a nucleon in the material of the metal plates of the chamber. The mesons resulting from the annihilation make the multiple tracks leading downward from the interaction vertex which are curved by the external magnetic field. This should be compared with the hydrogen bubble chamber pictures in Plates VIII–XII which are easier to interpret because the interaction takes place on a free proton.
(*Photo: CERN*)

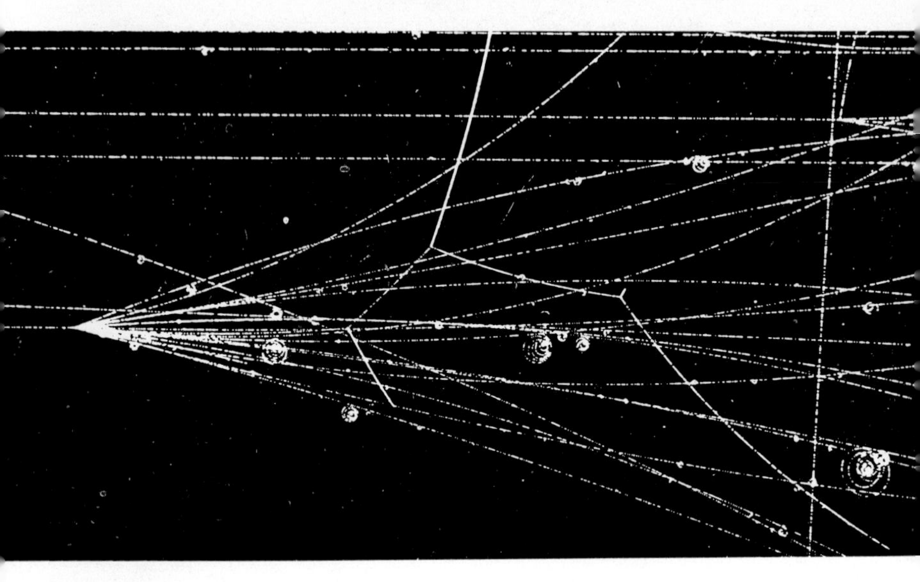

A hydrogen bubble chamber picture of a collision between a 16 GeV negative pion and a proton in the chamber, which has produced eight positive and nine negative pions.

$$\pi^- + p^+ \longrightarrow p^+ + 8\pi^+ + 9\pi^-.$$

One pion has made three successive bounces from other protons in the liquid. The sudden changes of direction and the heavy tracks of the recoiling protons are clearly visible. (*Photo: CERN*) decaying into two pions. This was accepted for about seven years to be the actual situation, but at the International High Energy Conference held at the Russian proton accelerator laboratory at Dubna in 1964 a group of American physicists announced that they had observed the forbidden two-pion mode for long lived *K*-mesons about once in every thousand decays.

In this experiment the 30 GeV proton beam of the Brookhaven machine was directed into a beryllium target. All charged particles were swept away using powerful magnetic fields and a channel about twenty metres long was constructed which could be traversed by neutral particles produced by the collision of the protons with the beryllium. At twenty metres all short-lived *K*-mesons have already decayed. The essential part of the experiment was an array of counters and spark chambers at the end of the channel to detect pairs of charged pions from the decay of long-lived neutral particles. Enough information was collected to determine the mass of the decaying particle by energy and momentum conservation. It was also possible to distinguish genuine two-pion decay from two chance pions which could arise from a neutral particle decaying into three pions.

The announcement of this detection of the 'forbidden' decay mode was met with considerable scepticism at the time, because it seemed such an ugly breakdown of a beautiful theory, and it is a very general experience that beautiful theories are usually right. However, since then this result has been confirmed in a number of experiments which are broadly similar in principle, but differ appreciably in detail from the original one described above. These experiments have also shown that there appears to be no alternative to the interpretation of the observation as a breakdown of CP-invariance.

Time's Arrow and Time Reversal

The significance of this important discovery depends on the usually assumed insensitivity of the primary physical laws to the direction of the flow of time. This requires rather careful consideration. The inevitable passing of time from one moment to the next is the physical effect of which we are most deeply conscious. If we cut out all normal contacts with the outside world by closing our eyes, blocking our ears and so forth, this innate feeling that time is passing is still very much with us, and moreover it has a very well-defined direction. Time marches on—not back! We have already considered playing a film from the wrong side, so that left and right are interchanged. Nothing very dramatic happens. But showing a film backwards so that the time ordering of events is reversed produces completely unbelievable situations. A diver emerges miraculously from a swimming-pool, lands neatly on the diving-board, his body bone dry the moment he leaves the water. A man regurgitates a banana, which bubbles out of his mouth, forms itself into a neat pillar of fruit on the stalk which he holds in his hand and is finally zipped back inside its skin. The idea is ridiculous, but in both cases, as far as physics is concerned, this inevitability of the direction of time's arrow arises from the complexity of the situation, not from the primary laws of physics. It is the purely statistical effect which has already been discussed at some length in Chapter 4 in connection with the stability of the proton.

If a system can exist in a very large number of different configurations an infinitesimal fraction of which are classified as ordered, any random shuffling process which shifts it from one configuration to another ensures by an overwhelming probability that it will be found in a disordered state. We now make the slight generalisation of this statement that, under these circumstances, if a closed system is seen to make a transition from an ordered to a disordered state, this determines the direction of time's arrow, with the disordered state occurring at the later time.

A man poised on a diving-board above a still pool constitutes a system with a certain quantity of gravitational energy in a highly organised form. The man may stand in various ways, but the situation is essentially unique. When he dives into the pool, this energy is dissipated in the motion of the water which can take on virtually innumerable different modes, and the almost uniquely ordered situation is replaced by one of enormous disorder. If at some instant after the diver has entered the water the motion of every particle in the diver and the water were suddenly reversed, a motion in the water would be set up in which a wave of pressure would converge on the diver and propel him out of the water, back on to the diving-board, much as a piece of wet soap jumps out of your hand when you squeeze it. Up to 1964 it was generally believed that this state of reversed motion would always be physically possible, and would not violate any primary physical law. Since this is the motion which one sees when time is run backwards, the fact that it is physically possible is referred to as invariance with respect to time reversal—or, in the jargon, T-reflection invariance. However, of all the innumerable possible configurations of the water in the pool only some infinitesimal fraction will have the property of ejecting the swimmer. The chance of such a rare configuration ever actually happening, as minor disturbances shuffle the water molecules through their various possible states of motion, is so near to absolute zero that it can safely be dismissed as impossible. Thus there are two general statements which one can make about the passage of time. The first refers to the primary laws and the precise specification of the motion of a system. It states that if, as expected, all primary interactions are invariant with respect to T-reflection, then the motion obtained by instantaneously reversing the velocities of all constituent particles of a physical system produces alternative motion which is also consistent with the primary laws. It is a physically possible motion. From this point of view the direction of time's arrow has no significance. The second statement is the statistical one made above, but let us repeat it. If the system is complicated, and its possible states can be classified into a relative few which are 'ordered' and almost infinitely many which are 'disordered', then the development in time will always be from the ordered to the disordered condition. The reversed motion, although formally allowed, is so improbable that it can be dismissed as impossible.

The close connection between the direction of time and the dissipation of ordered motion into disorder can be seen by replacing the diver by a ball bouncing on a floor. In this case the relevant disordered motion is the heat, or random motion of the atoms in the ball and floor, produced by the bouncing, which takes the place of the waves produced by the diver in the pool. If this dissipation effect is large, the ball rapidly loses its ordered energy. The heights of successive bounces get rapidly smaller as the ordered bouncing motion is converted into heat. In this case a film of the reversed motion, showing the ball apparently gaining energy from the floor, would be immediately detected by the experienced eye as going backwards in time. However, if the ball and floor are made of hard steel, and the dissipation of energy is reduced to a minimum, the atoms keep their places, order is maintained, the ball bounces almost to the same height each time. A time-reversed film would then be barely distinguishable from the correctly time-ordered motion. Thus the physical basis of the relentless onward flow of time, of which we are so deeply conscious, is apparently statistical in nature arising from the random effects of large numbers.

Time Reversal and CPT Invariance

The relationship between the primary physical laws to the possibility of reversing the order of events is quite different. We are looking for an explanation of the quantitative physical aspects of the world on a construction kit basis and expect in the end to find that it is made of a limited number of simple objects (like particles) interacting with each other in simple ways. The emphasis here is very much on the word simple, because anything complicated implies structure and demands further analysis in terms of yet simpler parts. If we believe that the present frontier of physics is in contact with the primary laws, it is already a surprise that they are showing a left–right complexity. But we would surely expect that at this level any interaction between particles which can be done up could be undone by simply reversing the motion.

This expectation appears to be true of gravitational, electromagnetic and Strong nuclear effects, but since the Weak interactions have been found sensitive to space reversal and particle–anti-particle reversal, we must consider carefully their invariance with respect to time reversal. So far there is no direct evidence of a breakdown, but there is an important general theorem. From the three separate operations of P-reflection, C-reflection and T-reflection we can make up a super CPT-reflection through which, given any system in motion, we construct another system by interchanging left and right, particle and anti-particle and finally reversing the direction of all motion including the direction of spin.

We have stated earlier that very general arguments from quantum mechanics and relativity lead one to expect the existence of anti-particles. The same considerations lead also to other conclusions about the relation between spin and the statistical properties of particles, the intrinsic parities of anti-particles and certain algebraic connections between processes which differ from each other by the replacement of some of the particles by their anti-particles. All these general predictions have proved correct, so one has by now considerable faith in this line of argument. Precisely the same considerations lead to the conclusion that primary interactions must be invariant with respect to CPT-reflection. This implies that given any physical system, then its super-reflection is also a physically possible system. Because of its excellent pedigree, this so-called CPT theorem is held in great respect.

Its consequences for at least some part of the Weak interactions are dramatic. If invariance with respect to CPT-reflections is satisfied, but that with respect to CP-reflection is broken, then invariance with respect to T-reflection must also be broken in some compensating way. We thus deduce that to some extent the primary Weak interaction is sensitive to the direction of time's arrow, and this very fundamental aspect of the physical world is lifted, at least in part, from the purely statistical aspect of the physics of complicated systems into the domain of the primary physical laws. There appears to be something in the basic mechanism of the Weak interaction which renders a reversal of some physically possible motion, not just statistically enormously improbable, but strictly and physically impossible—a profound development at a very critical point in the subject.

6

Patterns of Matter

Quarks and Unitary Groups

We now leave the curious complexities of the Weak interaction with regard to space and time, and return to the Strong interaction with its three conserved charges. Clearly the discovery of the conservation of baryon charge and hypercharge, together with the conservation of electric charge, puts a number of constraints on possible strong interaction processes, but the result is essentially a classification. The major step from classification to theory, which has been accomplished in the last few years, stems from work by some Japanese physicists which was presented at the International Conference at the University of Rochester, USA, in 1959. The paper was read in a final 'crack-pot' session in which theoreticians were invited to run free with their latest and wildest conjectures. 99% of such work leads nowhere, and this particular piece made very little impact at the time. It was certainly not recognised as being by far the most important contribution to the conference.

The implication of this Japanese work, as it was later interpreted, is made most convincing in the following way. The hadron states listed in Table 4 (page 75) fall into two groups of eight, distinguished from each other by the baryon number, the spin and, very roughly, by the mass. By this we mean simply that the baryons, with spin $\frac{1}{2}h$, are all roughly similar in mass to the nucleons (see Table 2, page 42), while the mesons of spin zero are about half as heavy. The pions with mass about one-seventh that of the nucleons are the only ones seriously out of line. It is evident from Table 4 that the distribution of electric charge and hypercharge within these octets is the same in the two cases. This can be most vividly displayed by making diagrams of the type shown in Fig. 3. The hypercharge, Y, is plotted on the vertical axis, and the other axis, Q, which denotes electric charge is taken at 30° to the horizontal. The electric charges and hypercharges of the particles can be interpreted as co-ordinates with respect to these axes. Thus, for example, K^+ which has electric charge one and hypercharge one has coordinates (+1, +1) and π^- with electric charge minus one and with hypercharge zero has coordinates (−1, 0). In this way

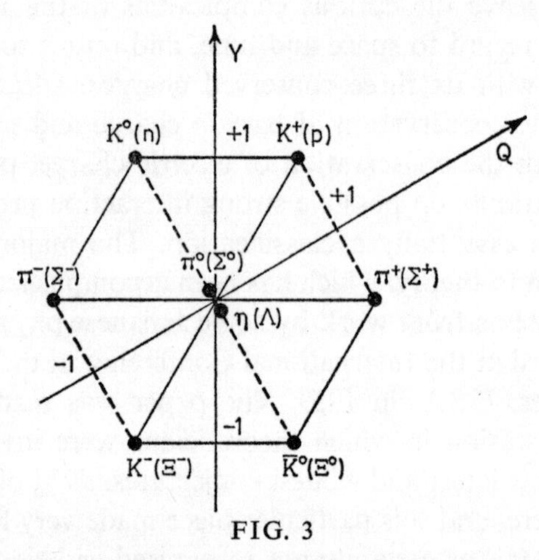

FIG. 3

The Q–Y plot of the eight spin zero mesons, and the eight spin $\frac{1}{2}h$ baryons (shown in brackets). Note from Table 2 (page 42) that particles which lie on the same horizontal lines in these plots have very nearly the same mass.

each particle in an octet determines a point on such a plot. The suggestive thing is that when so displayed each octet forms a regular hexagon with two particles at the centre. These elegant patterns are sometimes referred to as 'the eightfold-way', which is just a numerical pun. In spirit this development would have given immense delight to the Pythagoreans.

It is beyond belief that regularities of this sort can arise by accident. They demand an explanation and it is natural to be influenced by historical analogy. The last time that anything of this nature appeared in physics was in experiments in atomic spectra, where regularities such as the Balmer series and the

Periodic Table were found experimentally before any theoretical explanation was forthcoming. In that case everything was beautifully explained through the structure of the atoms in terms of protons and electrons. It was realised, after it had all been worked out, that a great deal of the regularity which had been observed was a direct consequence of the fact that the electrons move under a central attraction to the nucleus, which does not depend on the orientation of the atom in space. We have already seen that this leads to the conservation of angular momentum, but it also has other consequences. Because of the invariance with respect to rotations, the probability distribution of the electrons must either be spherically symmetric or, if it is not, the existence of one unsymmetric state implies the existence of others obtained from the first, by changing the orientation. The family of states which can be generated from each other in this way form a system which, taken as a whole, is spherically symmetric. For a hydrogen atom in an external magnetic field, which fixes a particular direction in space, these differently oriented but otherwise similar atomic states give rise to a multiplet of approximately equal mass values for the atom. These are reminiscent of the mass multiplets seen in the Q–Y plots of the hadrons. Less directly these families of states are responsible for the periodic grouping in the Periodic Table. For example, the closed shells of charge of the noble gases referred to in Chapter 1, arise when an atom has just enough electrons to fill one such set of states related to each other through rotation symmetry.

With this analogy in mind it was natural to ask what is the simplest sub-structure to the sub-nuclear goo which will lead to the hexagonal patterns? The answer is that you require three particles a, b, c which have been called *quarks*. Their charges are such as to form another regular pattern on a Q–Y plot, as shown in Fig. 4. These quarks automatically imply three anti-quarks $\bar{a}$, $\bar{b}$ and $\bar{c}$ with all charges reversed in sign. There are three of them because there are three types of charge. (The name is a reference to a remark in *Finnegan's Wake*: 'Three quarks for Mister Mark'—a rather desperate in-joke which may convince some that high energy physicists are both literate and numerate!) In addition to assuming the quarks—analogous to the electrons and protons in atomic structure—it is necessary to assume that the Strong nuclear forces, which operate between them are insensitive to which particular quarks are present. Given a physical system with a certain quark content, one must assume that alternative systems obtained by replacing the quarks by other quarks are also physically possible. This quark shuffling process is defined mathematically as operations of the

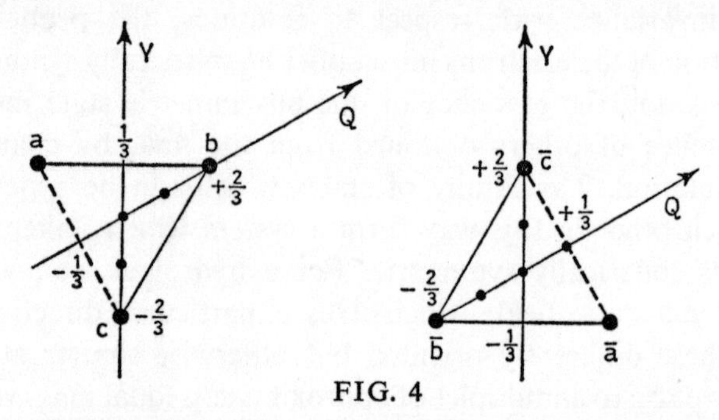

FIG. 4

The Q–Y plot of the basic SU(3) quark triplet (a, b, c) and the anti-quark triplet, ($\bar{a}$, $\bar{b}$, $\bar{c}$). The scale is the natural one to use for the hadrons and to produce both the meson and baryon patterns, the quark charges must be assumed to be fractions of those on the hadrons.

group SU(3)—the 3 referring to the number of quarks and SU to Special Unitary matrix transformations—and the precise statement of insensitivity of the quark forces to shuffling is that they are invariant with respect to these transformations. We have already seen that there is a close connection between invariance and conservation laws. Having found the conservation laws of the three types of charge, we are now introducing the invariance principle which is related to them.

Patterns of Quarks and the Omega Minus (Ω^-)

As in the case of the atom and its invariance with respect to rotations, invariance with respect to quark shuffling gives rise, not only to conservation laws, but also to multiplets. Let us take the simplest example. The baryon charge of quarks and anti-quarks must be equal and opposite, so a quark–anti-quark pair has baryon charge zero. Suppose that the combination $\bar{a}b$ forms a bound system, then any other quark–antiquark pair must also form a stable system by our shuffling rule. Since there are three quarks and three anti-quarks, there are nine such combinations. These split into a single symmetric quantum mechanical combination, $\bar{a}a + \bar{b}b + \bar{c}c$, which under shuffling goes back into itself, and eight other combinations, any one of which by shuffling implies all the others. The charges are just additive so once one has specified the charges on the quarks one has determined the charges on the pairs. These must be chosen so that the eight combinations reproduce exactly the eightfold hexagonal pattern as shown in Fig. 5.

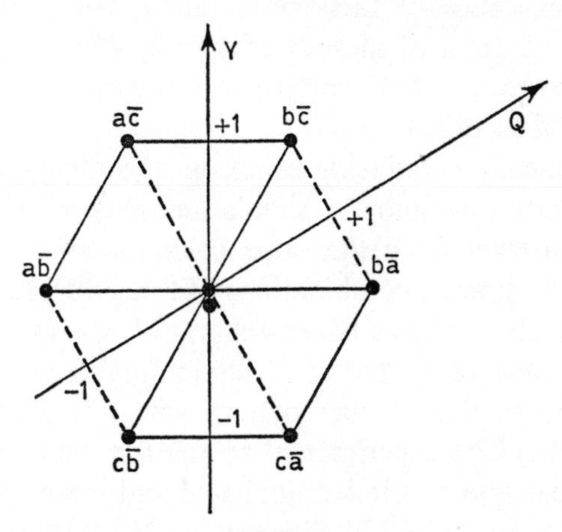

FIG. 5

The combination of quark–anti-quark pairs which gives the eightfold hexagonal pattern of observed mesons. The total charge of any combination such as *ac* is just the sum of the charges on the separate quark components, so the structure of this figure follows directly from the quark charges indicated in Figure 4.

The problem is rather tightly knit because from the same basic material we wish to produce not only the meson octet, but also the baryons. These can be constructed from special symmetry combinations of three quarks. Since the baryon charge of this three-quark combination must be one, this immediately implies that the baryon charge of the quark itself is one-third. The other charges also appear on a scale which is down from the normal one by a factor of three. The electric charges are $-\frac{1}{3}$, $\frac{2}{3}$ and $-\frac{1}{3}$, and hypercharges $\frac{1}{3}$, $\frac{1}{3}$ and $-\frac{2}{3}$. This is very strange. One of the few things which has survived all the discoveries of the last fifty years is that particles always seem to appear with electric charges which are simple integer multiples of the charge on the electron. We are being led to primary constituents of matter which do not follow this well-established pattern. However, let us not be too easily put off. When the theory was first proposed in this form the η-Meson was not known, so the meson octet was not complete. The theory started to attract general interest in 1961 when a variety of bubble-chamber experiments established the existence of this meson along with a new SU(3) singlet and an octet of mesons of spin h, which made up yet another hexagonal Q–Y pattern. However, many physicists still remained sceptical.

The completely convincing check on the theory came from the consideration of another simple multiplet structure, which one can construct from the same basic material. This is the decuplet of symmetric combinations of three quarks, illustrated in Fig. 6, which also has a baryon charge of one. (Not too much should be made of it, but it is fascinating how close these diagrams are to the number pattern which so impressed the Pythagoreans.) The experimental search for the allowed levels of sub-nuclear goo continued unabated and it was pointed out at the conference at CERN, Geneva, in 1962 that nine particles, five of which were only recently discovered, fitted neatly into the top three rows of the figure. On this basis the existence of the tenth particle was predicted. This is called Ω^- and its physical properties can be read off straight from the diagram. It has electric charge of minus one and hypercharge of minus two. Its spin is determined by the other members of the multiplet to be $\frac{3}{2}h$ and the mass is similarly prescribed. From this information

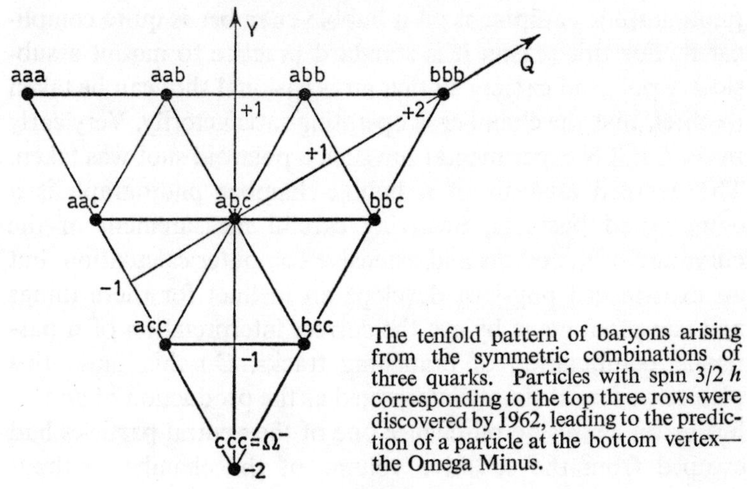

The tenfold pattern of baryons arising from the symmetric combinations of three quarks. Particles with spin 3/2 *h* corresponding to the top three rows were discovered by 1962, leading to the prediction of a particle at the bottom vertex—the Omega Minus.

FIG. 6

and the three charge conservational laws, it could be predicted that it should be produced in the process

$$K^-{}_{-1} + p_1{}^+ \longrightarrow \Omega^-{}_{-2} + K_1{}^+ + K_1{}^\circ,$$

implying an experiment with a high energy beam of *K*-mesons directed into a hydrogen bubble chamber. The possible decay schemes of Ω^- via the Weak interactions, subject to the rule that hypercharge changes by only one unit, are

$$\Omega^-{}_{-2} \longrightarrow \Xi^-{}_{-1} + \pi^\circ{}_0,$$
$$\Omega^-{}_{-2} \longrightarrow \Xi^\circ{}_{-1} + \pi^-{}_0,$$

or

$$\Omega^-{}_{-2} \longrightarrow \Lambda^\circ{}_0 + K^-{}_{-1}.$$

The mean life could be estimated to be about 10^{-10} seconds. The experimental requirements of a beam of *K*-mesons involved the sophisticated use of the full facilities of a 30 GeV proton accelerator. This was not achieved for another two years, but in 1964 both the laboratories of CERN and Brookhaven set up the proposed experiment and tension started to mount.

I can only tell the rest of the story as I experienced it. The full photographic equipment on a bubble chamber is quite complicated. For this reason it is standard practice to mount a subsidiary polaroid camera so that an occasional shot can be taken to check that the chamber is operating satisfactorily. Very early in the CERN experimental run such a polaroid shot was taken. The detailed analysis of a bubble-chamber photograph is a complicated business, involving careful measurement of the curvature of the tracks and extensive computer calculation, but an experienced physicist develops an instinct for these things and can often guess by eye the correct interpretation of a particular combination of branching tracks. On this basis this picture was most simply interpreted as the production of an Ω^-, but by an agonising mischance one of the neutral particles had escaped from the sensitive volume of the chamber without decaying into charged particles. This loss of information made it completely impossible to confirm the conjecture by careful measurement, but it aroused expectation and a copy of this photograph was pinned on a door at Imperial College by a colleague who was collaborating in the CERN experiment. That same day we had occasion to telephone Professor Bernardini at CERN about some quite unrelated business. On the telephone Bernardini made some remarks about the discovery of the Ω^- which were misunderstood at first, because we thought that he was referring to the inconclusive polaroid photograph. It was soon realised, however, that a message had just been received at CERN to the effect that the Brookhaven group, in the USA which was several weeks ahead of CERN with their experiment, had got an absolutely clear, fully measured up example of Ω^- production.

By such bush telegraph methods the news spread quickly from one laboratory to the next and a wave of excitement was felt by high energy physicists all over the world. Once again the apple had landed by Newton's head and a new primary physical law had been established.

The historic picture of the first established Ω^- is shown in Plate XI(a) and interpreted in Plate XI(b). The chain of events is

$$K^- + (p^+) \longrightarrow \Omega^- + K^+ + K^\circ$$

$$\hookrightarrow \Xi^\circ + \pi^-$$

$$\hookrightarrow \Lambda^\circ + \pi^\circ$$

$$\hookrightarrow \gamma \quad + \quad \gamma$$

$$\hookrightarrow e^+ + e^- \qquad \hookrightarrow e^+ + e^-$$

$$\hookrightarrow \pi^- + p^+$$

The most amazing thing about this photograph is that both the photons γ from the decay of π° have created electron-positron pairs. These made visible tracks which supply information essential for the unambiguous interpretation of the picture. The odds against this double-pair creation in the chamber are about $10^8 : 1$. On the basis of this it has been suggested that two important conclusions can be drawn. Firstly, that the Strong nuclear interactions are invariant with respect to SU(3). Secondly, God must be an American!

There are by now many more photographs of Ω^-. A very clear example showing an alternative decay mode is reproduced in Plate XII. In this picture only one neutral particle is involved, and this is shown up quite clearly by its decay into charged particles. The process is

$$K^- + (p^+) \longrightarrow \Omega^- + K^+ + K^+ + \pi^-$$

$$\hookrightarrow \Lambda^\circ + K^-$$

$$\hookrightarrow p^+ + \pi^-.$$

The foreshortened view along the K^- track shows the kink where the Ω^- decayed. It is also clear that the fast (almost straight) proton track from the decay of the Λ° points back, as it should, to this kink.

The discovery of Ω^- and the establishing of the SU(3) invariance for the Strong nuclear interactions must rank as one of the greatest achievements of human endeavour in the quarter century since the end of the Second World War. It is an extension of the domain of fundamental physics in the direct tradition of Newton and Maxwell and is an example of progress in scientific analysis in the absolutely classic manner. A period of intensely fruitful experimental activity gave rise to a situation which seemed theoretically chaotic and bewildering. Out of this apparent chaos a new general theoretical notion was born, which seemed to clarify the existing situation, but which initially had nothing sufficiently startling or unique about it to make it completely convincing. The crucial step was the prediction, on the basis of these ideas, of a new physical effect which was very precisely defined and absolutely specific to this particular line of thought. Finally came the experimental confirmation, verifying the prediction in every detail.

The story of Ω^- is also exceptional in that it is the first major scientific discovery to have involved the human race in a cooperative effort on a world scale. The detailed information on the multiplicity of hadrons, the beginnings of which are shown in Table 2, was based primarily on the big proton accelerator laboratories in the USA, western Europe and the USSR, but involved large numbers of physicists from all over the world. The idea of SU(3) invariance was started by Japanese theoreticians working in the University of Hiroshima—of all places! The correct identification of the formal multiplets of the group theory with the observed particle multiplets from the experiments was proposed independently in London and California. The crucial test of the theory was put forward at an international conference in Geneva; the experiment was successfully completed in Brookhaven, Long Island, USA.

In the aspirations of those engaged in scientific research this is the stuff of which dreams are made.

Sub-Nuclear Periodic Table

It is perfectly clear that the establishment of SU(3) invariance is an enormous step forward in Strong interaction hadron physics, but it still leaves a great many questions unanswered. The most obvious consequence of this new insight is that the discovery of one more hadronic particle is no longer of much significance. A 'new elementary particle' does not now make headlines in the popular press. It has become evident that there is an essentially infinite number of hadrons, just as there is an infinite number of possible excited states of atomic hydrogen or, in principle, an infinite number of different types of atom. In each case it is the least massive and therefore longest living states which are the most important. It is also quite clear that it makes no sense to treat all the hadrons as 'elementary' and, as in the case of atomic energy levels, their significance lies in the patterns which they make collectively, not in individual cases. As has already been suggested the discovery of the eightfold-way is in this respect very closely analogous to the finding of the Periodic Table of the chemical elements. By the early part of the 19th century the chemists had discovered that all matter is made up of about a hundred different atoms—or elements—two words both of which imply a basic unsplitable unit. This was an enormous simplification of the amazing variety of forms in which matter presents itself to us in everyday experience, but was still too complicated to be acceptable as a complete explanation. The next major step forward was the discovery by Mendeleev in 1869 that these atoms could be arranged in groups, and that the chemical properties of any given element could be predicted from its position in the group. When this Periodic Table was first proposed a large number of elements predicted by the scheme were not known and were subsequently discovered on the basis of their chemical properties, as predicted from their position in their group. The discovery of the element hafnium as late as 1922 is a famous example. The clarification of the significance of the hundred-odd known hadrons brought about by the eightfold-way—or SU(3) invariance—is clearly of a very similar nature, and the discovery of Ω^- closely parallels the discovery of the rare chemical elements such as hafnium.

The next step in the understanding of atomic structure was the development of the quantum theory of the possible energy levels of electrons moving under the influence of the electrical attraction of an atomic nucleus. This was conjectured by Bohr, but given first in a complete and self-consistent form by Schrödinger. It explained regularities such as the Balmer series concerning the frequencies of photons emitted as an atom makes a transition from one allowed energy level to another. As discussed in more detail in Chapter 1, and mentioned above, it also explained the grouping of elements in the Periodic Table, which is closely related to the multiplets of states that follow from the invariance of the whole system with respect to rotations.

The corresponding step in hadron physics would be the development of an analogous theory, which would determine the allowed masses of the hadrons and relate them to their other properties. The particles are specified at the moment by their mass, spin and intrinsic parity. The mass determines the rest energy which is associated with invariance with respect to displacements in time. The spin is similarly related to rotations, and the parity to mirror reflections. All these three properties are thus intimately connected with space-time and with the various transformations one can make in space-time without altering the form of the theory. Quite independent of these is the newly discovered invariance of the Strong interaction with respect to the shuffling of quarks—invariance with respect to the group of transformations of SU(3). This is the underlying physical property which leads to the conservation of the three types of charge, and to the formation of the eightfold-way multiplets with their patterns of Q–Y charge values. Since the shuffling transformations do not involve any changes in the space-time coordinates, there is no connection between the charge multiplets, which arise from SU(3), and the spin multiplets coming from invariance with respect to rotations in space. In particular the SU(3) theory leads naturally to the baryons appearing in charge multiplets with eight or ten members, but does not explain why the former should have spin $\frac{1}{2}h$ and the latter spin $\frac{3}{2}h$.

Larger Patterns, Larger Groups

The exciting suggestion was made in 1964 that a spin of $\frac{1}{2}h$ should be attributed to each of the three quarks, and that the invariance should be extended to shuffling which includes the reversing of the direction of the spin component together with the replacing of one type of quark by another. Since each of the three quarks has two possible spin orientations—'up' and 'down'—there are altogether six possible quark states, and the group of transformations which shuffles these six states indiscriminately is SU(6). The procedure for constructing multiplets is exactly as before, except that now spins are involved and are consequently specified in the final structures. If we again consider the mesons to be quark–anti-quark pairs, there is one completely symmetric combination of zero spin, zero electric charge and zero hypercharge, which goes into itself under shuffling. Then there is a large multiplet consisting of a unitary octet of spin h, another of spin zero and a unitary singlet of spin h. Particles of zero spin have only one possible spin state, but each particle of spin h has three possible states, corresponding to the three allowed orientations of the spin axis. Counting all the different spin states and charge states separately, this leads to a super-multiplet of thirty-five states:

$$(8 \times 3) + (8 \times 1) + (1 \times 3) = 35.$$

The existence of any one of these implies all the others, by shuffling. These thirty-five related states and the unrelated singlet state correctly describe all the well-established mesons with masses up to about the mass of the proton, giving the spins and parities in the observed combinations with SU(3) multiplet structure. The model is equally successful with the baryons. The generalisation to SU(6) of the symmetry combination of three quarks which makes up the octet in SU(3) is a baryon supermultiplet consisting of an octet with spin $\frac{1}{2}h$ and a decuplet of spin $\frac{3}{2}h$, exactly as observed. This supermultiplet of fifty-six states incorporates all the baryons which appear in the discussion of the previous chapter from the nucleons right through to the Ω^-, and again correlates correctly the spin and parity properties with the patterns of charges.

It thus begins to appear that all the properties of the sub-nuclear goo, which is the basis of Yukawa's model of the strong interactions and the particles arising from it, can be explained by assuming that it is made of quarks and anti-quarks with spin of $\frac{1}{2}h$. We shall consider below how literally one can take this statement, but let us for the moment accept it at face value. One can now modify the approach to the general problem of hadron masses to ask more specifically what is the physical mechanism which determines the masses of the heavier hadronic states. This is the most clearly formulated problem in this field at the moment, and the one on which most direct progress has been made in recent years. In terms of our quark model there are two obvious alternatives. One is to build up the mass by adding extra quark–anti-quark pairs to the combinations already considered, which one can do without altering the baryon charge. The more massive multiplets are then generated by SU(6) shuffling on all the quark constituents of the more complicated quark structure. Thus, for example, larger meson multiplets with zero baryon charge can be generated by SU(6) shuffling of two, rather than one, quark–anti-quark pair. Since on this picture spin and charge are treated on a very similar footing, these multiplets have the general property that the maximum hypercharge occurring in the multiplet increases roughly in proportion with the maximum spin. An alternative possibility is that the higher mass states are not obtained by adding quark–anti-quark pairs, but by the original quark combinations acquiring additional energy, and hence also mass, from a bodily rotation. On this basis the total spins of the hadrons would be observed to increase as the mass increased, with no corresponding increase in the complexity of the charge patterns. To decide between these possibilities is a difficult question, which cannot be solved by a single experiment, but only by a steady accumulation of data by many different experimental groups. The number of established hadrons has more than doubled since the discovery of Ω^- in 1964, and a picture is starting to emerge which seems to favour the latter view, though the complete answer may well be some combination of the two possibilities. It is too early to draw any definite conclusions. An answer to this general problem will probably depend on an extension of experimental facilities, both in terms of available beam energy and the precision with which observations can be made.

The Duality of Production and Interaction

We have been concentrating on the spectrum of hadrons as they are observed in production processes in high energy collision. But we must remember that the pions were originally predicted by Yukawa as the lightest lumps of goo which can be exchanged by two interacting nucleons in a glancing collision. In more direct head-on collisions the heavier hadrons are exchanged. In this alternative capacity the hadrons do not appear as end products of collision, but provide the mechanism through which the primary hadrons—the particles in the beam and the target—interact with each other. In this indirect way the hadrons determine the dependence on energy and angle of the disposition of the primary hadrons, following collisions in which no new particles are produced. The fact that the hadrons themselves provide the mechanism through which other hadrons interact has been called 'duality' and the self-consistency conditions implied by this dual role has been much studied in recent years.

An interpretation of high energy collision experiments in terms of the implied hadron exchanges also tends to favour the idea that, at least in part, quark combinations acquire extra energy through rotation rather than by the accumulation of extra quark–anti-quark pairs. On this view a particular quark configuration, such as that leading to the thirty-fivefold multiplet of mesons, should form a whole family of hadrons in which the charge-spin patterns keep repeating with increasing total angular momentum, and consequently increasing energy, due to the rotation of the whole quark system. These, so-called, Regge families can be shown to act together in their other capacity of providing a mechanism for the hadron interaction and, when exchanged, give rise to a characteristic energy and angular dependence in those collisions in which the primary particles bounce off each other without the production of secondary particles.

The further analysis of the Strong interaction is thus the subject of a double-pronged attack. There is the direct approach in which one studies the masses, spins and charges of the hadrons which appear in those collisions in which new particles are produced. There is the indirect approach in which one concentrates on the dependence on energy and angle of the primary hadrons following those collisions in which no new particles are produced. The latter information has to be interpreted in terms of exchange of hadrons by the Yukawa mechanism. The dual roles in which the hadrons appear, either directly as end-products in collisions, or indirectly as objects which can be exchanged, thereby providing the mechanism of interaction in the collision, have to be made consistent with each other.

Hierarchy of Interactions and the Coupling Currents

We are still a long way from the mass formula which would be the analogue of the Schrödinger equation for the hydrogen levels. In this respect one should emphasise that SU(3)—and still more SU(6)—are only approximate invariance properties of the Strong interaction. If they were exact, all the hadrons in a given multiplet would have exactly the same mass, and at this level of approximation all that could emerge would be a formula for the mean mass of each SU(3) multiplet. In actual fact the observed masses within a multiplet differ by about one-tenth of a proton mass between elements which are related by the shuffling of *c*-quarks. However, if they are related by the interchange of only *a* and *b*, they have the same *Y*-value and very nearly the same mass (see Figure 3). The implications of this are that the interaction which we have previously designated as Strong must be subdivided into really Strong, which is invariant with respect to SU(3), and a Medium Strong interaction, which is invariant for the shuffling of *a*, *b* quarks but not of *c*, *a* or *c*, *b*. More physically one may think of the *a* and *b* quarks as having essentially the same mass, and the *c*-quark as being appreciably different. The *a* and *b* quarks should be expected to have a small mass difference of a few thousandths of a proton mass due to the difference in the energies required to assemble their differing electric charges. It is reasonable to suppose that this is the cause of the relatively small mass differences observed between multiplet members with the same *Y*-value, which differ from each other only by *a* and *b* exchanges.

We are thus led to a hierarchy of interactions, whose invariance properties increase with the strength of the interaction. The strongest is the Strong nuclear interaction which is invariant for space-time displacements, rotations and reflections, and for all SU(3) shufflings of the *a*, *b* and *c* quarks. The Medium Strong is similar, but is only invariant for *a*, *b* shufflings. Next in strength is the electromagnetic interaction which is not invariant for any quark shuffling, but still conserves all three types of charge, *Q*, *Y* and *B*, and is invariant with respect to all the space-time transformations. Then comes the Weak interaction which conserves electric and baryon charge, but not hypercharge and also fails to be invariant with respect to space-reflection. Finally, in the sub-nuclear domain is the Super-Weak interaction for which time-reversal invariance also breaks down. This interaction, however, is invariant for space-time displacements and rotations so that energy, momentum and angular momentum are conserved, as also are baryon and electric charge. These appear to be absolute conservation laws which are satisfied by all interactions. Not included in this discussion is the gravitational interaction, because it is so weak that it appears to play no significant role in the sub-nuclear domain. It is generally assumed to be invariant with respect to all the space-time transformations, time-reversals and space-reflection, but in view of the apparent correlation between breakdown of invariance and weakness of interaction, it might turn out to have surprising properties.

There is no known explanation for this correlation between the strength of the interaction and the variety of its invariance properties. Nor, indeed, is there even a remote understanding of why there should be the various distinct types of interaction, widely separated by their intrinsic strengths. However, an interesting connecting link between them has been remarked, which also stems from SU(3) invariance.

To the extent that one can ignore all effects except the Strong nuclear interactions, all phenomena are exactly invariant with respect to SU(3) transformations, and there are then actually *eight* conserved charges. Corresponding to each of the charges there is a current density just as there is an electric current density related to the flow of electric charge. Of the eight SU(3) charges, only two can be observed simultaneously because of mutual disturbances between their measurements—similar in principle to the mutual disturbances between measurements of position and momentum of an electron—but all eight currents can be defined. It is a remarkable fact that the electric current of the hadrons, which determines their electromagnetic interactions, is a linear combination of two of the SU(3) currents which arise in connection with purely Strong effects. The Weak interactions of the hadrons can also be expressed in terms of four other currents, and these also turn out to be four more of those coming from SU(3). Thus the electromagnetic and Weak interactions of the hadrons can both be expressed in terms of hadronic currents, which arise in the first place in the context of the Strong interactions only. At the moment this just stands as a fascinating experimental fact, but it is a rather clear indication that the various interactions, which when classified by their strengths and invariance properties appear so disjointed, may well have some common origin and become amenable to some more general theory.

There are, of course, many other problems such as the mechanism for the breakdown of time-reversal invariance and the question of whether there is a specifically leptonic goo which mediates the Weak interactions, in the same way that the exchange of hadronic goo mediates the Strong. These are general questions which can already be formulated on the basis of present knowledge. If we extrapolate from the experience of the last twenty-five years, the most interesting developments in this field during the rest of this century will be ones which are totally unforeseen at present, but which will emerge as beams of higher energies become available and make possible the study of physical situations never previously produced under laboratory conditions.

Do Quarks Exist?

We have deliberately avoided raising until now the obvious question of whether quarks exist, or if they are just a convenient mathematical fiction for calculating, and to some extent visualising, the properties of hadrons. This question is obviously one of enormous interest, but it highlights a much more general point, which is crucial to the whole question of a fundamental basis to the physical aspects of the Universe.

To establish the existence of quarks, in the sense in which physicists use the word, it must be possible to observe them as separate entities, for example, through the tracks which they should make in a bubble chamber. These would exhibit the mass and momentum of the quark through the curvature of the tracks in a magnetic field and the restrictions following from energy and momentum conservation. It should also be possible to detect their charge through the density of bubbles along the track. For particles travelling near the speed of light this density depends on the square of the charge, so the tracks of a particle of charge $\frac{1}{3}$ should be an order of magnitude fainter than those of a normally charged particle and should be readily detectable. If hadrons are made of quarks, then quarks should be produced in hadronic collisions provided the process is consistent with all the relevant conservation laws. Because of their fractional charges there is no possibility of producing them one at a time, since a final state involving particles with an electric charge one-third of the electron unit cannot possibly arise in charge-conserving collisions between normal particles with integer charges. However, quarks could be produced in threes or as quark–anti-quark pairs. The latter have all total charges zero—electric, baryon and hyper—so that as far as these conservation laws are concerned a quark pair can be produced from very general initial conditions.

Since the difference between the quark charges is integer on the normal scale, the heaviest quark, the *c*-quark, say, of charge $-\frac{1}{3}$ can decay by ordinary Weak interactions into the *b* quark, of charge $\frac{2}{3}$, with the emission of leptons

$$c^{-\frac{1}{3}} \longrightarrow b^{\frac{2}{3}} + e^{-1} + \bar{\nu}_e{}^{\circ}.$$

This is just the same β-decay mechanism by which a neutron decays into a proton. However, the lightest quark must be stable in surroundings consisting of ordinary integer charge particles, because there is no way consistent with charge conservation that it can get rid of its fractional charge. Thus, once produced, one would expect that the presence of quarks could be detected.

Extensive searches have been made for quarks during the last five years in nuclear collisions using the proton accelerators and in the much higher energy events occurring in cosmic rays. No conclusive evidence has been found. The frequency of collision in cosmic rays is so low that not much can be deduced from that negative evidence, but the absence of quarks at machine energies forces one to the conclusion that they must be at least five times as heavy as protons. In this case the energy required to produce quark pairs is not available with proton accelerators in operation at present and their absence is simply explained on the basis of energy conservation. If they fail to materialise at still higher energies, this can always be explained by simply raising the lower limit of the mass. As the mass becomes heavier their significance as real particles becomes less and less until in the limit of infinite mass, they are reduced finally to a mathematical fiction, introduced simply to help define the shuffling processes of SU(3). These could have been defined directly in terms of the members of the eightfold multiplet, but the mathematics is then far less transparent. The validity of the SU(3) patterns does not depend on the existence of quarks, but the lower limit of five proton masses is sufficient for the important general point we wish to make.

Open-ended Physics

If protons and pions are to be interpreted as made of such massive quarks, then we are confronted with the paradoxical situation of a whole which weighs only one-fifteenth (for the pion it is one-seventieth) of the sum of the masses of its parts. It is clear that this is a very important development and allows a complete escape from the corner into which physics appeared to have been driven by the analytic—or atomistic—approach, as it has been interpreted during the last three hundred years. If you stamp on your watch and it disintegrates into cogs and springs, it is reasonable to suppose that these are the constituents out of which it was made, and to examine its working and its structure in the light of this information. This is an essentially limited problem, and the objects one finds in this way are limited to these which were in the watch to start with. However, if you kick a dog and it barks, to deduce that a dog is made of barks would be unimaginative and misleading. In high energy physics the dogs have started to bark, and the very notion of quarks as a substructure for hadronic matter has turned fundamental physics from what looked thirty years ago like an almost closed book into a completely open-ended subject.

Actually nothing new is happening. It is a matter of degree and all we are witnessing is the obvious logical consequence of Einstein's realisation that mass is a form of energy which is interchangeable with other forms. A hydrogen atom weighs less than a free electron and a free proton by about one part in a thousand million. The effect is so small that one can still think of a hydrogen atom as being made of an electron and a proton, in just the same sense that a machine is made of parts and matter is made of chemical elements. The parts and the elements are identified to a large extent by their mass. Thus in a car we distinguish between major components, such as the engine block and the back axle, and little bits like nuts and bolts. In chemistry the identification of the elements by weight is complete, and it is a basic working rule of the subject that the sum of the masses is equal to the mass of the whole.

In the properties of atomic nuclei the interchangeability of mass and energy starts to show itself significantly for the first time, but this has not yet induced a corresponding change of outlook. The mass of a typical nucleus is about 1% lighter than the mass of the constituent nucleons. This mass defect is a measure of how tightly the nucleons are held together by the Strong nuclear forces and is exactly equivalent to the energy required to disintegrate the nucleus into its constituent parts. This is the nuclear phenomenon which has had the most violent effect on the human condition, but in the physics of nuclear structure it is still possible to think of a nucleus as made of nucleons in a quite conventional sense. In the theory of nuclear structure, the properties of a nucleus follow from the properties and interactions of the nucleons of which it is composed and, as in the case of a watch, the nature of the constituent parts determines, almost by definition, the boundaries of the subject.

This limitation applies to any physical system so long as the kinetic energy involved in the process of division and analysis is completely negligible compared with the rest energy of the system being analysed. Normally this can be taken for granted. The watchmaker dismantles a watch with delicate instruments appropriate to the job and would not dream of going at it with hammer and chisel or by slamming two watches together. But in hadron physics we have for the first time a complete reversal of the situation. There is no delicate probe available and to bang the hadrons together is the only possibility. In experiments using 30 GeV proton accelerators, the kinetic energy in the equivalent head-on collision between beam and target particles is already about six times greater than the proton rest energy. The exciting thing is that this has turned out to be a creative process which has completely upset the atomist notion that there is a limit to divisibility and a logical end to physics. The products of such a collision depend not so much on the identity of the colliding particles as on the amount of kinetic energy which is fed in. The end products of these collision, like the barks from the dog, are only constituents of the original particles in some very loose sense, and the subject has switched under our noses from a study of the properties of hadrons to the much wider study of the physics of very high energy densities. As the energy is increased there is no logical limit to the richness of what may be uncovered, any more than in astronomy is there any limit to what may be found as we look deeper into space. At 250 GeV we may break into the realm of quark dynamics in which the properties of quarks as individual entities can be studied. This will be a completely new field of physics. At even higher energies new layers of reality may be uncovered, and there is no reason why these should be any less rich in the variety of their physical properties than chemistry, which is the physics of atoms, or modern high energy physics, which is predominantly concerned with the properties of hadrons. In Nature these exotic processes may now be taking place in extremely high energy cosmic ray collisions or at the centre of the densest stars.

Quark Fields or Nuclear Democracy

Correspondingly, the structure of a physical system such as a proton is now seen to be no absolute thing, as assumed in the atomistic approach, but a relative concept which depends on the energies involved and the particular properties which are being studied. At low energies protons behave dynamically like structureless points. As the energy of collisions is increased to a level at which pion production is energetically possible, the proton may be regarded as having a 'bare' proton core surrounded by a cloud of mesons—the extra mass being cancelled by the binding energy. As the energy is increased still further the whole variety of hadrons start to emerge from collisions. If one persists in the atomistic notion that any system is made of the bits which are found when it is knocked to pieces, a more elaborate concept emerges of each hadron being a complicated conglomerate of all other hadrons, no one of which is any more fundamental than any of the others. This is known as 'sub-nuclear democracy' and is certainly a valid picture within its limits. In apparent contrast to this we have the surprising experimental fact that the static properties of the hadrons other than their mass, such as their charges, parities and spins, can be very beautifully explained on the assumption that they are entirely made up of very simple configurations of small numbers of quarks. If the quarks have finite mass these too must eventually emerge among the debris of sufficiently high energy collisions. From the dynamic point of view these too must be included among the constituents of hadrons, even if they are not regarded as the only constituents.

The quark picture and the hadronic picture are not in fact contradictory because, if hadrons are made of quarks, then hadrons made of hadrons are also made of quarks. The group theoretic argument allows for the possibility that in addition to the few quarks that determine the Q–Y patterns, each hadron contains a symmetric residue of any number of quark–antiquark pairs, which transforms into itself under shuffling. But although consistent with each other, the two pictures are clearly redundant and each may be useful in its appropriate context. The least that has become clear is that the structure revealed by analysis, in these circumstances, is not well-defined and there is no substance in the argument that physics must ultimately have a stop in constituent parts too simple to be analysed further.

Of course an open-ended subject is more interesting than a closed one, but one can have too much of a good thing, and physics proceeds by the discovery of unifying syntheses which incorporate a wide variety of phenomena in terms of a few simple concepts. This procedure amounts to at least the partial closure of an open end. Since the analysis of matter in terms of particles, specified primarily by their mass, has proved a bottomless pit, one is very conscious of the problem of finding some alternative concepts which will give a simple unified picture of hadronic matter, at least in the highly idealised approximation in which only the Strong SU(3) symmetric interaction is operative. One possibility is the relativistic quantum quark field. This bears much the same relation to the quark as electric and magnetic fields do to the photon in the quantum theory of radiation. The energy density of radiation can be expressed in terms of these fields and it is conceivable that hadronic energy density, which we have seen is the central physical property, can be simply expressed in terms of the quark field, interacting with itself. This could remain true, even in the limit in which the quark mass is taken to be infinite, and the notion of quark particles becomes a pure abstraction. The hadrons would then appear as singularities in this field. Conceptually this may be interpreted as going over from the study of knots to the study of string. The knots may be of great variety and their analysis in terms of simpler knots might not lead very far, but once it is realised that they are all made out of one piece of string, one is back again to something conceptually simple. This is an attractive idea, but it is rather empty at the moment because with very strong self-coupling of the quark field there is no known method of actually calculating what it would imply in terms of the masses of the particles and their behaviour in collisions.

7

Prediction

Mass, Length and Time

The changes in outlook brought about by the discoveries of sub-nuclear physics have naturally had an influence on the way physicists think about their subject and on the generally accepted notion of its structure and significance. The new attitudes have been more drastically affected by the development of atomic physics than by the nuclear and sub-nuclear physics, which is our main theme. However, this is the level at which the subject makes its broadest impact on thought in other fields, so we will attempt a general statement of the present position. It might seem more logical to have done this at the beginning of our discussion, but we have preferred to follow the standard practice of physicists which is first to make some progress and then to sit back and assess what has been achieved. This attitude is possible because physics is an essentially pragmatic subject.

We gave a definition of the scope of physics in Chapter 1, but let us now be more precise. It is an attempt to find a unified description of the information we receive about the outside world through our senses, with particular emphasis, as pointed out in Chapter 1, on isolating those aspects which underlie what is causal, repeatable and predictable in everyday experience. The outside world is assumed to have an independent existence only to the extent that the things which give rise to sense impressions are found to be the same for everybody. If, of two men in a room, one sees a chair standing in the corner and the other does not, it is presumed that one or other of them needs medical attention. Since it is totally impractical to study the entire Universe in one go, it is often convenient to consider

ydrogen bubble chamber photograph of anti-proton which annihilates with a ton in the liquid hydrogen into three sons. *(Photo: CERN)*

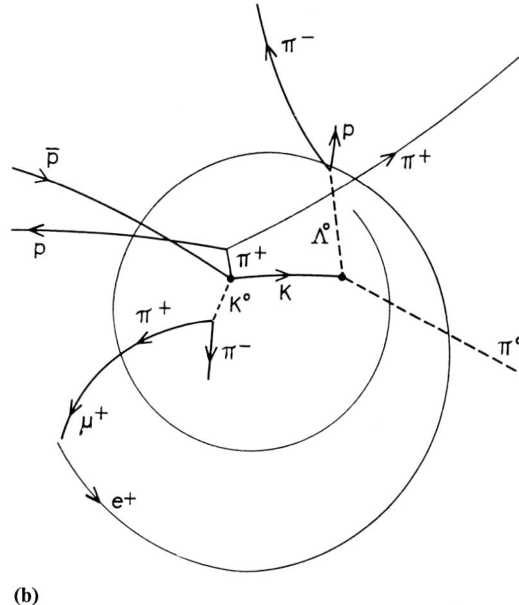

(b)

The processes photographed are indicated n the key plan:

$$\bar{p}^- + (p^+) \longrightarrow K^\circ + K^- + \pi^+$$

$$\pi + (p^+) \longrightarrow p^+ + \pi^+$$

$$K^- + (p^+) \longrightarrow \Lambda^\circ + \pi^\circ$$

$$\hookrightarrow p^+ + \pi^-$$

$$\rightarrow \pi^+ + \pi^-$$

$$\hookrightarrow \mu^+ + \nu_\mu{}^\circ$$

$$\hookrightarrow e^+ + \nu_e{}^\circ + \nu_\mu{}^\circ.$$

Stationary target protons in the liquid which do not give tracks are shown in brackets. Strong interactions are denoted by heavy arrows, Weak decays by light arrows.

X

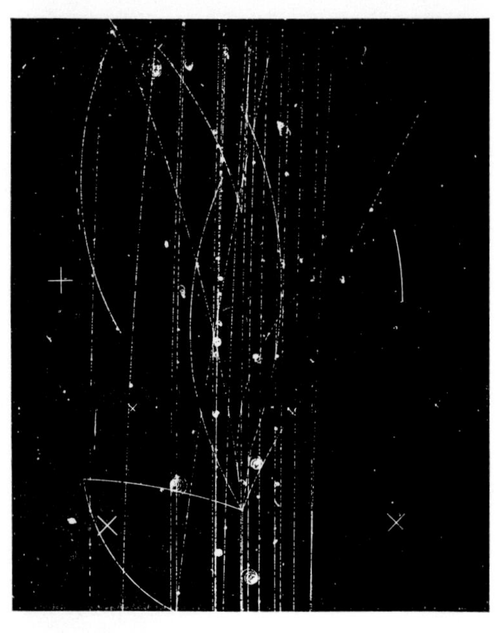

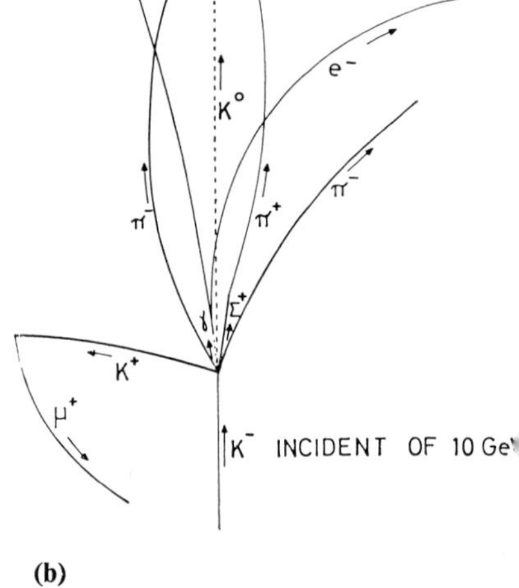

(a)

A hydrogen bubble chamber photograph of events produced by 10 GeV negative *K*-mesons entering from the bottom of the picture.

(*Photo: CERN*)

(b)

The π° does not live long enough to le visible gap. It decays almost at the poi production into two photons γ, one of gives rise to the visible electron-positron 'closed systems' which are isolated and often deliberately simplified subsections of the world. We have frequently made use of this notion in the previous chapters. It is assumed that these closed systems can be protected from all outside influence to any desired extent. Thus the Sun and the planets form a closed system to a very good approximation, but one may also profitably treat the Earth and the Sun, or a single hydrogen atom, as deliberately simplified closed systems.

In addition to the outside world we must include observers who receive the sense impressions, but who also figure in physics in a very specialised way. From the whole variety of the physical qualities perceived by the senses, a rigid selection is made of those aspects which may be described in quantitative terms on the basis of well-defined operations of measurement. There turn out to be just three such aspects, which may be taken to be *mass*, *length* and *time*. All three, including the units in which they are measured, are defined by experimental procedures which ultimately imply direct comparison with very specific physical standards. Length and mass were originally defined by the metre rule and the kilogramme mass, two pieces of platinum carefully preserved in the International Bureau of Weights and Measures at Sèvres in France. The unit of time, the second, was defined as a definite fraction of the time taken by the Earth to orbit the Sun in the year 1900! All other physical quantities, such as velocity, momentum, energy and electric charge, together with a consistent set of units in which they can be measured, can be constructed from the three given above, which are referred to as 'dimensions'. The choice of mass, length and time is not unique, but these, until very recently, appeared to be conceptually the simplest, mass being the general quantitative property of matter, and length and time being similarly related to the space-time continum in which matter is seen—or sensed—to move. The standards for comparison, which also define the units of measurement of these quantities, are even more arbitrary and recently more readily accessible and universal standards have been introduced, directly related to Planck's constant and the velocity of light.* However, the essential point remains that sense impressions are codified through measurement by numbers. The word 'observer' is to be understood as a shorthand for this whole process of making measurements, including the physical apparatus which it entails. The human mind only comes in at the very last stage, converting the position of a pointer on the scale into a number and recording the nature of the measurement and the units used. Thus, for example, if the measurement were of a velocity the result could be in the form 2·5 metres/second; if it were an energy one might obtain 17·35 kilogramme metres^2/second^2.

The Power of Mathematics

If physics were to stop at this point it would be the most tedious collection of information imaginable. The next step is the most adventurous and the least justifiable *a priori*. It is at this point that creative genius in the sciences is so similar to the greatest work in the arts. It is here that in terms of human achievement Newton takes his place with Boticelli, Shakespeare and Beethoven. It has proved possible by an amazing combination of inspiration and guesswork to extrapolate from the necessarily approximate numerical relations, which may be found between a finite number of observations, to mathematical equations between symbols standing for whole classes of measurement. In this way one establishes a correspondence between the raw data of physics and branches of pure mathematics. The latter are logical structures—free creations of the human mind—

* Traditionally one starts with physically defined units of length, time and mass.

The unit of length (metre rule) can be replaced by the velocity of light, which provides a universal standard. The unit of length is then the light-second, which is the distance travelled by light in one second.

One can go still further in this direction. Instead of making a direct physical definition of the second, one can take Planck's constant h as the universal unit of angular momentum. Given some unit of mass M (which can be the kilogramme at the International Bureau or, more conveniently, something universal like the mass of a free proton) the unit of length is then h/Mc, that of time h/Mc^2, energy Mc^2, momentum Mc and (electric charge)^2 is measured in units of hc. Modern definitions of length and time are closely related to these natural, universal units.

following directly from some set of self-consistent axioms and are independent of sense impressions. The distinction between mathematics and physics is important and is shown very clearly through the dimensions. Because mathematics is logically independent of the physical world the entities appearing in it do not have dimensions, in the sense in which we are using the word. The quantities occurring in physics, on the other hand, are meaningless unless the dimensions are specified. As described in Chapter 3, a primitive check on the possible validity of the equations of mathematical physics—as opposed to pure mathematics—is that the former must relate physical quantities of the same dimensions. The subject of mathematics is of course enormously wider than that of mathematical physics, and there are whole fields of mathematics, such as the more abstract realms of geometry and group theory, which bear no relation to the physical world. The validity of a piece of mathematics depends on the rigour of the argument. In contrast the validity of a theory in mathematical physics is judged by its success. All the relevant known data must fit the formulae, and the hallmark of success is the inference of new and previously unsuspected physical effects, by mathematical reasoning, on the basis of a particular theory. The discoveries of Ω^- through the group theory of SU(3) and of the planets Neptune and Pluto from the discrepancies between the observed orbits of the other planets and the predictions of the theory of gravity, are outstanding examples. There is nothing absolute about the validity of a theory since there is always the possibility of new data being found, which do not fit it. The significance of a theory depends on the economy of its concepts and the generality of its conclusions. The importance of the transition from physical data to mathematical theory is that it unleashes on the problem the extraordinary power of mathematical reasoning to establish inter-connections of remarkable subtlety which could not be found by other means and imparts to physics the beautiful precision of mathematical statements. In particular it provides a very precise definition of a predetermined system.

Newton, Laplace and Determinism

All these features are clearly illustrated in the research leading to the discovery of Ω^- and the physics of planetary motion. In the latter case the vital role of measurement was stressed by Galileo, who made the observations rendered technically possible by the invention of the telescope. A prodigious amount of experimental work was carried out by Kepler, who extracted from the data three empirical rules:

(i) that the planetary orbits are ellipses with the Sun as focus;
(ii) that the line joining the Sun to a planet sweeps out equal areas in equal times (this is just the conservation of angular momentum);
(iii) that for each planet the square of the orbital period is proportional to the cube of the mean distance from the sun.

Newton's theory, in the equivalent but more elegant form in which it was later cast by Hamilton, then shows that if the Sun and a planet are treated as an approximate closed system, their entire motion can be derived from the expression for the energy, including the gravitation binding energy, written in terms of the masses, positions and momenta of the two bodies. Kepler's laws are included, incidentally, among the consequences of this detailed analysis. The planetary motion is given by differential equations which describe the motion of *any* system containing two bodies which is dominated by the gravitational interaction between them. The motion of a particular system is determined by the masses (presumed constant) and the positions and momenta at any instant. From this information the positions and momenta at any time in either the future or the past can be calculated. The solution, of course, ceases to be physically significant as one works back to the time of the creation of the planets, since then the Sun and a single planet could no longer be reasonably treated as a closed system. However, within such obvious limitations the theory provides an absolutely clear mathematical description of a predictable system, and a precise statement of the information required to make a firm prediction. In this way in his *Principia Mathematica* published in 1687, Newton introduced the notion of a completely deterministic closed system. It was Laplace born in 1749, just over twenty years after Newton's death, who made the sweeping generalisation of this notion to the entire Universe. In his *Theories analytique des Probabilités*, published 1820, he wrote:

> An intelligence knowing at a given instant of time all forces acting in nature as well as the momentary positions of all things of which the universe consists, would be able to comprehend the motions of the largest bodies of the world and those of the lightest atoms in one single formula, provided his intellect were sufficiently powerful to subject all data to analysis; to him nothing would be uncertain both, past and future would be present to his eyes.

This enormously powerful idea has greatly affected the general attitudes of Western man, producing, as it does, a mixture of wonder and horror. The wonder follows obviously from the magnificence of the conception; the horror because it seems to leave little place for life and none for free will. Man feels instinctively that he can influence the course of history which Laplace maintains could already have been written. Moreover, it is this power which makes him a man. All thinkers before Thales and many after him assumed that generalisations of this type of influence would be the driving forces of the Universe. By contrast, in Laplace's vision these forces of soul and mind seem to have got squeezed out altogether.

Laplace's statement is clearly much more scientifically based than the straightforward fatalism of Omar Khayyam:

> Yea, the first Morning of Creation wrote
> What the last Dawn of Reckoning shall read

—in that he makes it clear that the prediction must be based on data (sense impressions), so that both the observer and the outside world are essentially involved. However, the idea of a deterministic Universe depends on two implicit assumptions, both of which have proved only partially correct, in a way that drastically restricts its validity.

Observations

Let us suppose that the world is made of atoms, as suggested by Laplace, but as understood in 1930. Suppose further that the nuclear structure can be ignored so that only the motions and positions of the electrons and nuclei need be considered. In Laplace's Universe the roles of observer and outside world are inextricably mixed up, so let us go back to the simpler notion of an observer and a closed system, which can be taken to be a single hydrogen atom. If the Laplace conjecture is to work for the whole Universe, it should certainly apply to this enormously simplified model.

At this point it is absolutely essential to remember that physics is about sense impressions. Our knowledge of the electron is limited by the extent to which it gives rise to sense impressions, from which we select certain quantitative measurements. Physics is about the interplay between the observer and the closed system. If the closed system were completely closed there would be no interaction between it and the observer—and no physics! To collect his information the observer must interact with the closed system, using probes. These are themselves subject to the laws of physics which he is trying to find, and we must generalise the notion of a closed system to one which is isolated from all outside influence except those necessary for the observer to make his observations. It is a crucial tacit assumption of classical physics—and this is the first of the assumptions referred to above—that the effects of the observer can be made vanishingly small, so the processes of measurement do not appreciably disturb the system observed. In these circumstances the observer is of no significance and the emphasis is entirely on the closed system. The importance of the observer and the dependence of observations on the physical properties of his probes was first appreciated through the theory of relativity. As we describe in Chapter 1, this shows that the finiteness and universality of the velocity of light makes it inevitable that the observed distances and time intervals between two events depend on the state of motion of the observer. However, the Principle of Relativity demands that the laws of mathematical physics should be of such a form that they are not significantly affected by this dependence, so this in itself does not upset Laplace's argument.

However, for the motion of electrons in atoms it has been found that the assumption that observations do not disturb the closed system is simply not true. The simplest probe the observer can use is radiation—or photons. There is an inherent fuzziness in measurements of position made with photons, which is comparable to their wavelength. To make precise measurements it is necessary to use photons of short wavelengths. But from the properties of photons described in Chapter 1, this implies photons of high frequency and consequently high energy and high momentum. It turns out that to make an accurate measurement of the position of an electron in an atom, it must be struck with photons of such high momentum that the resulting recoil appreciably alters the momentum of the electron in a random way. Measurements of the momentum affect the position in a complementary manner. Thus classical physics deals with information about the outside world of the type we are accustomed to receive through our eyes which is collected without affecting what is seen. (If there are disturbances, corrections can be made for them.) In atomic physics we deal with the sort of world which would be sensed by intelligent beings endowed only with a clumsy sense of touch, who are liable to knock over and appreciably disturb in an unpredictable way any small object with which they come in contact. In these circumstances it is not surprising that, in a mathematical theory of the interaction between an observer and a closed system, one has to distinguish between the operations of measurement and the numerical results. In such cases the assumptions of classical physics are not valid, and one is forced to go over to quantum physics in which operations of measurement are represented by mathematical operators.

Quantum Mechanics

Newton had to develop for himself the pure mathematics of differential calculus, which he needed to calculate the planetary orbits. The discoverers of quantum physics were luckier to the extent that the algebra of operators—either matrices or differential operators—had already been worked out. This is different from the familiar algebra of functions in a very obvious way. If s and t stand for numbers, and st for their product, the order in which they are taken is not significant. In general

$$st = ts.$$

However, if $\hat{s}$ stands, for example, for the operation 'one step forward' and $\hat{t}$ for 'turn right', then $\hat{s}\hat{t}$ represents the operations carried out in one order and $\hat{t}\hat{s}$ represents the opposite order. It is very easy to check by direct trial that these two combined operations do not have the same effect. Symbolically, this is written,

$$\hat{s}\hat{t} \neq \hat{t}\hat{s}.$$

In classical physics we deal directly with the numerical data, and can introduce ordinary algebraic variables x and p standing for the position and momentum of a particle. For these it is always true that

$$px = xp. \qquad (1)$$

In quantum physics we must introduce operators $\hat{x}$ and $\hat{p}$ to stand for the corresponding operations of measurement; $\hat{p}\hat{x}$ is then connected with the two successive observations in one order, and $\hat{x}\hat{p}$ is similarly related to the operations in the reverse order. Because of the inevitable mutual disturbances, these combined operations are not equivalent, so

$$\hat{p}\hat{x} \neq \hat{x}\hat{p}.$$

The difference between the two depends on the mutual disturbances which one measurement has on the result of the other and has the dimensions of Planck's constant h. Since h is a measure of when quantum effects become important, it is reasonable that the difference between the operators should be simply related to it. The correct expression is*

$$\hat{x}\hat{p} - \hat{p}\hat{x} = ih. \quad (2)$$

In the mathematical theory each operator is associated with a set of numbers, called eigenvalues. In the physical theory these numbers are identified with the possible results of the corresponding measurements. Thus x and p, the possible values of position and momentum, are the eigenvalues of the operators $\hat{x}$ and $\hat{p}$, which represent the operations of making measurements. The operator which represents the measurement of energy is related to the operators $\hat{x}$ and $\hat{p}$ in the same way that the classical expression for the energy is related to the variables x and p. The Schrödinger equation for the energy levels of a hydrogen atom is just the equation to determine the eigenvalues of the energy operator. The formula, which is a special case of a completely consistent general theory, agrees with that obtained earlier by Bohr using his brilliantly conceived, but arbitrary and self-contradictory rules tacked on to the classical theory of Newton and Maxwell. The fact that only a certain discrete set of energy levels is allowed, corresponding to the angular momentum increasing in integer multiples of h, now appears quite naturally from the formalism. The mysterious factor i, in the relation (2) above for the operators, is required to ensure that the results of measurement predicted by the theory are all real numbers, not involving such a factor.

Quantum mechanics turns out to be self-consistent in a very important way. We have seen that the experimentally observed break-up of radiation energy into discrete photons imposes certain limitations on the use of radiation as a probe for the observation of small, closed systems. Quantum mechanics has been constructed to take account of these limitations and provides a general and consistent method for calculating the possible energy levels of any closed quantum system. When this theory is applied to radiation considering it not as a probe, but as a

* We have used the notation $i = \sqrt{-1}$.

closed system, it is found that the possible values for the energy are just those to be expected for an assembly of photons. Thus, in a very direct way, quantum mechanics gives back the break-up of radiation into photons, which originally motivated the theory, but which was initially incorporated in a very indirect manner.

The state of a closed classical system, such as an orbiting planet, can be specified at any instant by giving its position x and momentum p. The mathematical operators, $\hat{x}$ and $\hat{p}$, operate on functions of either x or p and in the physical theory of a quantum system, such as an electron in an atom, these functions specify the state of the system on which the operations of observation are made. A state function—sometimes called wave function—does not specify precisely the position and momentum of the electron since these cannot be simultaneously observed. Instead, its square modulus defines a certain probability distribution for these variables which is as detailed as possible, subject to the mutual disturbances between the two types of observation. It thus quantifies an optimum situation, and completely describes an electron in as much detail as it can be sensed by an observer. The product of the uncertainties in position and momentum described by the state function is approximately equal to h, which is a consequence of the statement (2) about the operators $\hat{x}$ and $\hat{p}$.

Quantum Predictions

The mathematical theory still has predictive power for closed systems for which quantum effects are important, but it is modified from the precise form it takes in classical theory by the random disturbances due to observations. At any instant the state of a closed system is specified by an optimum set of measurements made at that instant. If no further measurements are made, the state is completely determined at some later time by differential equations, which are qualitatively similar to those which determine the time development of a classical system. The new state specifies a new probability distribution of the observable properties, which is related to the old one, but in which the probabilities tend to be more diffuse. Thus, if at the initial time the quantum state approximates to a single classical state, at a later time the probability is spread over a wider range of classical situations. If new observations are made at this later time, the results will be consistent with these causal predictions of the state, based on the earlier measurements. However, the new measurements provide new information and introduce new random disturbances. The new observations force the system to make a discontinuous jump into a new state, which specifies the new relationship between the observer and the physical system.

Thus, if the system consists of an atomic electron, the observer may choose to make very infrequent observations, in which case the state of the electron develops predictably in time but contains rather vague information. Alternatively the observer may make very frequent observations. Then the state at any given time contains more precise information, but the development in time is much more erratic, since a greater random element has been fed in by the extra measurements.

We are so accustomed to think in classical terms that it is tempting to ask what the 'real' electron is doing during the intervals between the observations on which the state function is based. At this question the physicist can only reply. 'I do not know. If you want an answer to that you must ask a philosopher. He won't know either, but he will probably tell you.' We repeat again that physics is about sense impressions, and the maximum amount of sense data we can have of the electron is summarised in the state function. To ask for information which cannot be incorporated in the state function is to ask for unattainable information, which takes us out of physics into metaphysics. However, the physicist can make a very definite and important negative statement. It is certainly not true that the electron actually follows a classical orbit with definite positions and momenta which, owing to our clumsiness and the limitations imposed by our probes, we have been unable to discover. When electrons are deflected one at a time from a crystal lattice, they form diffraction patterns just like a train of light waves through a diffraction grating. This observed diffraction pattern arises only because each electron takes a variety of possible paths through the crystal, with various probabilities. If it actually took any particular one of these paths the observed diffraction patterns would not form. If additional measurements are made to determine which path it takes, the pattern is destroyed.

In spite of this strange behaviour, which simply does not fit into the frame of thought we have developed from the study of macroscopic systems, the quantum theory goes over quite smoothly into the classical theory in the appropriate limit. This can be seen by considering what happens when, conceptually, the masses of the electron and the atomic nucleus are allowed to increase till h becomes negligible. Then the disturbance accompanying observations also become negligible, and precise measurements can be made simultaneously of all observable properties such as position and momentum. Formally this is expressed by replacing h by zero in the right-hand side of relations such as (2) above. It then becomes identical with (1) and there is no need to introduce operators into the formalism. The random element disappears from the quantum prediction of how the system develops in time, even when it is subjected to frequent observation. In these circumstances the quantum predictions coincide exactly with classical predictions. Whenever h is negligible, the state function of a particle with mass ceases to be significant, and it is the position and momentum with precise values which survive in the classical limit. However, for photons, with no mass, the particle aspects fade out and the state function, or wave function which determines the frequency and wavelength, is of dominating importance. It is for this reason that photons were first found as waves in the classical physics of macroscopic objects.

Classical physics, including its predictions of the time development of a system, is an approximate form of the more generally valid quantum physics. The classical approximation is valid to the extent that Planck's constant *h*, and consequently the disturbances accompanying measurements, are negligible in comparison with the properties of the closed system under consideration.

This is obviously the case for the motion of astronomical bodies, and space rocketry provides an ideal opportunity for man to display his mastery of the physical situation. Classical mechanics does not apply to the motion of electrons in atoms or in molecules where chemical bonding is essentially a quantum effect. However, in most situations in electronics the uncertainties of quantum mechanics are completely negligible, and electrons can then be treated as though they were billiard balls. This is true, for example, of the electrons responsible for the image on a television screen, where positions are specified to about 10^{-4} metres and tens of volts correspond to momenta of about 10^{-26} in MKS units. The product, which is of the dimensions of Planck's constant is 10^{-30} MKS compared to which h (10^{-34} MKS) is quite negligible.

Let us get back, finally, to the very simple model of a Laplacian Universe consisting of one hydrogen atom and an 'intelligence' or observer, whose sense impressions and powers of measurement are subject to the same fundamental limitations as those of an actual observer. Since Laplace's concept is put forward as a logical consequence of physical laws, it has no basis unless this restriction is imposed. To make it more interesting let us suppose that at the beginning of time the 'intelligence' has observed the electron to be in a state in which it is approaching the proton with momentum and position specified as precisely as is allowed by the quantum conditions. Provided the 'intelligence' then leaves the 'Universe' alone the whole future *state* is 'present to his eyes', but this only determines the various probabilities that after a definite time the electron has been deflected into some particular direction, or captured into an atomic orbital with the emission of photons. At the end of time, on the day of judgement, the 'intelligence' can make further observations to find out which of these alternatives has actually taken place. If he chooses to collect further information at some intermediate times, he necessarily affects the course of events in a completely random way. Since the actual Universe certainly contains vast numbers of electrons in essentially quantum situations, there is no basis left in physics for the vision of iron determinism in which the orbit of every particle is specified in detail on the basis of a complete classical description of the initial conditions.

The Basic Variables

But, as we have said, there is another implicit assumption in the Laplace conception. This can be dealt with quite briefly since it is closely connected with the general implications of sub-nuclear physics discussed at length in the previous chapter, although not directly in this connection.

Even if one ignores the restrictive conditions imposed by quantum mechanics, it is clear that the deterministic concept only makes sense if it is possible to specify what information is required in order to make a prediction. As long as matter, measured by its mass, is regarded as the basic stuff of the Universe, it is natural to suppose that the information could be given in terms of the positions and momenta of the ultimate particles, as proposed by Laplace. However, we have seen that the theory of relativity replaces mass by energy as the base quantity in physics, and the logical case collapses for what we have called the 'limit to divisibility'. A hydrogen atom can be described in terms of an electron and a proton. But the proton itself has structure, and we have not yet found an economic set of observable properties from which it can be defined, and on the basis of which prognostications can be made. To an extent this is a problem peculiar to this age. In rather general terms the basic question facing high energy physics at the moment is to establish the elementary set of observables appropriate to sub-nuclear physics. This problem can no doubt be solved. The concept of quarks is a step in this direction, and some generalisation of this idea may provide the ultimate variables for the description of

Nature. It may be that however much energy is concentrated in a given volume no new layers of reality emerge. But, as stressed in Chapter 6, we have no reason to assume this, and our vision of the Universe itself becomes open-ended if it is impossible to specify the observables in terms of which all its physical properties can be defined.

We see that the two major general developments in physics in the 20th century, relativity and quantum mechanics, have both dealt severe blows to the deterministic view of the Universe, which was made so popular and so plausible by the physics of the 18th and 19th centuries. It is, of course, impossible to disprove a metaphysical belief that everything which happens is predetermined, but one can have good or bad reasons for thinking that it might be. The 19th-century physicists appeared to have produced rather good reasons. But the theory of relativity has shown that there may be no basis of facts from which the physical determinism could start. Quantum mechanics implies that even if there were, to collect this information would itself vastly disrupt the course of subsequent events in a completely random way. As we have said, none of this disproves a rigidly deterministic Universe, but 20th-century physics cannot be cited as a basis for a belief in it.

Statistical Predictions

We have been discussing this question in a somewhat formal way, because it is of such interest as a matter of principle. But for practical purposes there is a much simpler limitation to the exact predictive power of physical theories which was already mentioned in Chapter 1. This is just the fact that we are usually confronted with systems in which the effective number of constituents is far greater than can be handled in practice by the direct application of primary laws. The problem of two gravitating bodies can be solved exactly, as shown by Newton. More than two bodies already call for approximations, and Laplace used great ingenuity with approximate methods to apply Newton's theory to the entire solar system. He may have been inspired by his success with an eight-body system to extrapolate the method, conceptually, to the whole Universe. But the extrapolation is completely unrealistic, since any macroscopic piece of matter contains millions of millions of millions of atoms. The only way forward is to make a virtue of necessity and actually exploit the presence of these large numbers. The method is that of Statistical Mechanics. It has already been discussed somewhat indirectly in Chapter 4 in connection with the stability of the proton, and very directly in Chapter 5 where we were concerned with the significance of the direction of time's arrow in relation to primary and statistical laws. The argument is so simple that it is surprising that it is so powerful. We have seen that any closed system may be found in conditions of relative order or disorder, which are classified according to the number of states in which they can occur. Here states are well-defined configurations of the system giving the maximum amount of information about the system consistent with the quantum conditions. The number of states in the disordered condition is usually not just greater, but enormously greater than in the ordered condition. To simplify the subsequent discussion we assume this to be the case. The crucial statement is that, if at some initial time the system is in an ordered condition, it can safely be predicted that in the course of time it will go over into the disordered condition and never get back again. All that is required is that there is some interaction which makes transitions between states possible. In making such predictions, the physicist is in the happy position of a gambler who can bet time and time again with the odds consistently and overwhelmingly in his favour.

A very simple example is a mixture of two gases contained in a given volume. In this case the states are well-defined motions of the gas atoms, determined by their positions and momenta at any instant. The interactions, which shuffle the system from one state to another, are the atomic collisions. If, at any instant, the gases are seen to be separated in opposite halves of the available volume (presumably because a partition separating them has

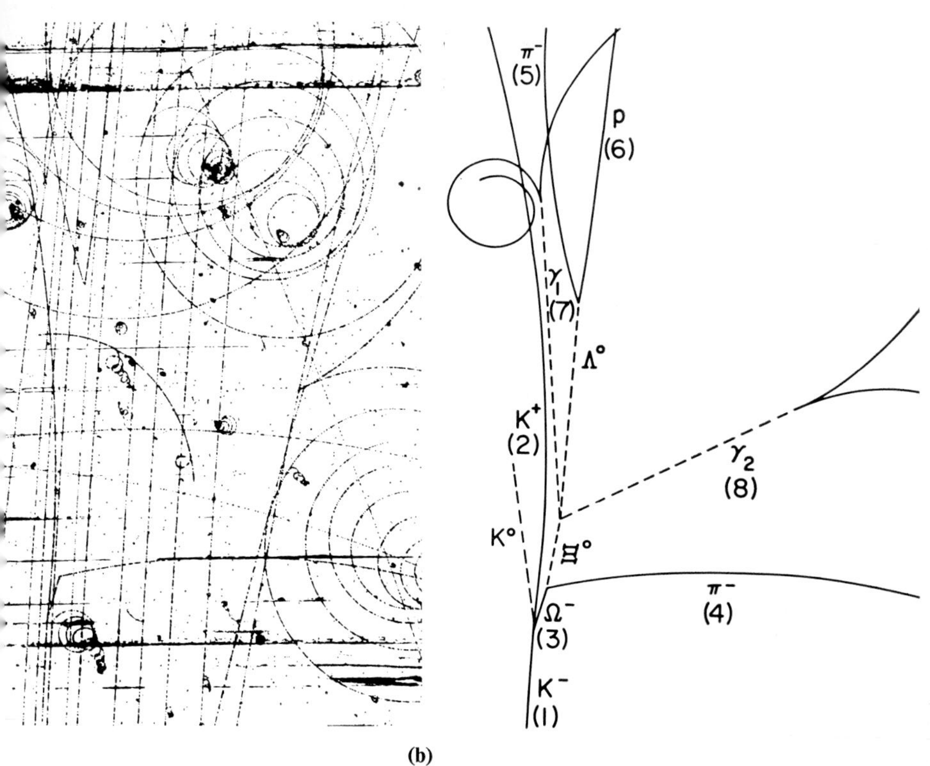

famous bubble chamber picture showing irst clearly identified example of Omega ıs. The negative *K*-mesons enter the picture the bottom.

(*Brookhaven National Laboratory*)

(b)

The sequence of events is:

$$K^- + (p^+) \longrightarrow \Omega^- + K^+ + K^\circ$$
$$\Omega^- \rightarrow \Xi^\circ + \pi^-$$
$$\Xi^\circ \rightarrow \Lambda^\circ + \pi^\circ$$
$$\pi^\circ \rightarrow \gamma + \gamma, \quad \gamma \rightarrow e^+ + e^-, \quad \gamma \rightarrow e^+ + e^-$$
$$\Lambda^\circ \rightarrow p^+ + \pi^-.$$

As in Plate X, the π° decays at its point of production into two photons, both of which have produced $e^+ + e^-$ pairs.

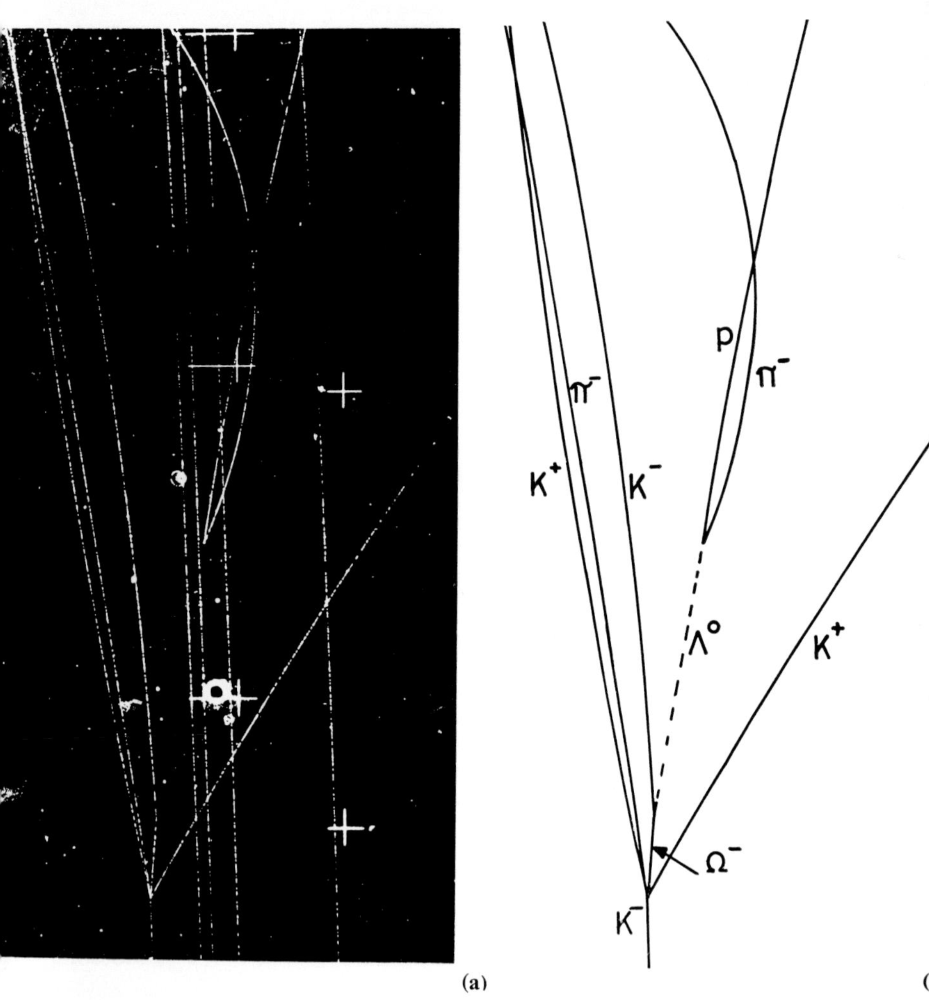

(a)

Another example of Ω^- production, in a collision between a negative K-meson and a proton in the liquid hydrogen of a bubble chamber.

$$K^- + (p^+) \longrightarrow \Omega^- + K^+ + K^+ + \pi^-$$
$$\Omega^- \rightarrow \Lambda^\circ + K^-$$
$$\Lambda^\circ \rightarrow p^+ + \pi^-$$

The foreshortened view along the K^- track shows clearly the kink where the Ω^- decayed. *(Photo: CERN)* just been removed) it is quite certain that the gases will expand and mix to fill the whole volume and, left to their own devices, will never again separate. This follows directly from the fact that the number of mixed states is enormously greater than the number of separated states. This in turn is a consequence of the large numbers involved, as may be seen by considering the opposite extreme in which each gas consists of only one or two atoms. In this case the prediction could clearly not be made, since the atoms would quite frequently separate by chance into the opposite halves of the container. Another interesting feature of this type of argument is that since it is itself statistical, it does not matter whether the unique states contain exact or statistical information. All that is required is that they can be counted. It is, in fact, somewhat simpler to set up statistical mechanics for quantum than for classical systems, because the counting of states for quantum systems is clearly defined. The vast majority of predictions in physics are based, at least in part, on this type of statistical argument.

Many of the following points have been made already, but we are now in a position to tie up a number of loose ends. It was pointed out in Chapter 4 in connection with the stability of the proton, that mass is a very highly ordered form of energy. If in a given system it is possible for part of the mass (rest energy) to convert to kinetic energy the number of states is always increased. This is because kinetic energy implies motion, which can be oriented in a great variety of different ways to give a great variety of distinguishable states. There is, therefore, a general tendency for matter to convert into the lightest possible configuration, the rest energy being transformed into the kinetic energy of photons and neutrinos. We have seen that the sort of detailed prediction for the Universe proposed by Laplace, even if it were valid in principle, would be totally impossible in practice. However, qualitative statistical predictions can certainly be made, and it is interesting that the general outline of cosmological development, given in Chapter 3, may be interpreted rather simply as the continuous increase of disorder, according to statistical mechanics, subject to the general restrictions of the primary laws. These laws determine which type of shuffling takes place between states and, through the strength of the relevant interaction, the speed at which it goes.

The Model Universe

To make the point it is sufficient to take a rough model in which the Universe is treated as a closed system of very large, but finite, volume for which the total energy and the total electric and baryon charges are conserved throughout its motion. On the basis of what we see of the Universe, we may suppose that at some early stage it consisted of a diffuse gas of hadrons and leptons of zero total electric charge but non-zero baryon charge at zero temperature. The baryon number and the total energy, which is the rest energy of the original particles, remain unchanged throughout the subsequent motion. The reduction of mass and the consequent conversion of rest energy into the kinetic energy of photons and neutrinos, which follow from the general statistical argument, would then go through the following stages.

(i) The hadrons would decay into protons, electrons, photons and neutrinos via the Weak and electromagnetic interactions; further degeneration in this direction being blocked by the conservation of baryon and electric charge.

(ii) The protons and electrons would then form hydrogen atoms through the electromagnetic attraction, which are lighter than the mass of the constituent particles by the mass equivalent of the binding energy. The total energy of the whole system would remain constant by the emission of more photons.

(iii) Then would follow the formation of galaxies, star-clusters and finally stars by condensation of the hydrogen gas under gravity, the mass being still further reduced by the gravitational binding energy. In this case a considerable fraction of the energy goes into heat at the centres of the stars.

(iv) This triggers off the conversion of hydrogen to helium

via Weak and Strong nuclear forces at the centre of the stars (as described in some detail in Chapter 3). This gives rise to a rapid conversion of mass into kinetic energy, since for the first time Strong nuclear forces play a role.

The process of nuclear burning within the star then continues with the formation of heavier elements up to iron, the surplus energy still escaping from the star as photons and neutrinos. All of this agrees with what has been observed for some time. However, if the star is big enough the gravitational forces get so overwhelming that the electric charge is squeezed out through the Weak interaction process closely related to normal β-decay,

$$e^- + p^+ \longrightarrow n^\circ + \nu_e^\circ.$$

The system becomes wildly unstable and the outer surface blasts off in a super Nova explosion, but the rump of the baryons then fuse together to form one lump of nuclear matter consisting entirely of neutrons—a neutron star. This is lighter than the sum of the masses of the original free baryons by the mass equivalent to the enormous amount of energy necessary to tear it apart against the tremendous forces of the nuclear and gravitational attractions working in unison. Although by this stage about one-fifth of the original mass has been turned to kinetic energy, the remaining matter in this form is enormously dense. Owing to the inward pull of the forces, a handful weighs at least a thousand million tons. These strange objects were predicted by nuclear physicists thirty years ago and would seem to have been discovered recently, first by radio astronomers, in the form of pulsars. The identification is not yet complete in detail, but the essential point is that at these densities the whole mass of the Sun would be concentrated in a volume of radius about ten kilometres. An object this size could have natural oscillations with the rapid frequency which is the typical and striking feature of the pulsar signals. In this way a very rough-and-ready application of statistical principles combined with our knowledge of the fundamental interactions, gives a general understanding of the way the Universe is developing.

The model is too rough to be taken much further. The assumption of a finite volume leads to a night sky ablaze with photons. Just to explain why the stars are seen against a dark background, we need to generalise the model to allow for the expansion of the Universe and the effects of general relativity.

Another problem is the question of the stability of large neutron stars. If the mass is very big, it is not clear whether the Strong hadronic interactions are capable of withstanding the enormous gravitational pressure produced at the centre of such objects. Astrophysics is a fascinating interplay of statistical effects and the basic interactions of Nature with the Weak and Strong nuclear interactions playing a very essential role. At the centre of large neutron stars we are led directly to the great unanswered questions of super-high energy physics and the structure of space-time under very extreme conditions considered in General Relativity theory. This takes us outside the scope of this volume. Our crude model of the Universe is sufficient to show how genuine physical predictions, even on a cosmic scale, can be made, and how these depend on a combination of statistics and the fundamental physics laws in which we are primarily interested here.

Of course, in more specific situations much more definite prediction can be made, which often contain a combination of statistical argument and the precise Newtonian form. This is well illustrated in electrical effects. Maxwell's equations are directly applicable to the flow of charge in empty space. The properties of a good conductor like copper can be understood on the basis of these equations and the existence of atoms through a whole series of quantum effects. The physical basis of the electrical properties are described in general terms in Chapter 1, which include the formation of the atomic nuclei into a regular lattice structure and the existence of a band of electrons which are free to travel through it. In addition to the quantum probabilities relating to the motion of individual electrons, there are statistical effects related to the large number of electrons involved. However, all the statistical complications are averaged out owing to the large numbers and the effect can be summed up in just one number, the conductivity. By a further direct use of Maxwell's equations this determines the current in, for example, a copper wire which results from a given voltage (Ohm's law). The practical equations which are used by the light and heavy electrical industry all follow in the same way in terms of relatively few similar parameters. These make up a deterministic scheme governing the flow of electrons, which does for macroscopic electrical systems what Newton's theory did for the motion of the planets.

Life and Thought

A question of outstanding current interest is the extent to which biological processes can be treated in the same way. On the one hand quantum mechanics and relativity have destroyed the supposedly scientific case for belief in a rigid determinism. This has left room in our understanding of the material Universe for thought processes generally, and for decision-making in particular. On the other hand there are the rapid advances in biophysics and biochemistry, including the unravelling of the genetic code and the analysis of the nervous system in terms of electrical circuits. It is clear that most aspects of the life process can be explained in physical terms. But the concepts of physics appear to become totally inadequate when we come to emotion, feeling and, particularly, to a sense of values. Between the electrical signals coming through the eye to the brain and our reaction to the vision of a tree in blossom on a fresh spring day, there is a vast gap which physics shows no signs of ever being able to bridge. This is just one example of the limitations of the scientific approach. We are confronted once again with the world of thought, mind and spirit, and are back to the old conflict between the Ionians and the Pythagoreans. In the language of Eddington's analogy of the ichthyologist, there is no doubt that the net of measurement, with which the physicist catches his quantitative information, has a finite mesh and a great deal of what is significant in life slips through it. Physics says a great deal about Truth, but nothing about Beauty, and is not concerned with moral judgements. It may even be that whatever it is that is peculiar to life and particular to thought lies outside the scope of physical concepts.

Even at the level at which the concepts of physics do provide an adequate basis for the description of life processes, something very strange is going on in living matter, which relates directly to the general physical view of the Universe developed above.

We have seen that an important general effect in the time development of inanimate physical systems is the random shuffling among the various states in which a system can exist, subject to the restrictions imposed by the general conservation laws. As time passes this shuffling always tends to take the system from relatively ordered condition to a relatively disordered one. However, within local regions of space-time there is always the possibility of the reverse process by which the degree of order is increased, provided it is balanced by a decrease of order in some neighbouring region. We refer to this local increase of order as sorting. This unlikely transition from a more probable to a less probable state can happen to some extent as a random fluctuation, but it is frequently true that the appearance of sorting is a good indication that life processes are at work. A single cell is already quite an elaborately ordered structure. A human being is, at very least, an assembly of chemicals constructed and maintained in a state of fantastically complicated organisation of quite unimaginable improbability. As soon as life is extinguished the processes of decomposition and decay, which are simple manifestations of increasing disorder, again take control of the situation.

Of course the appearance of sorting, by itself, is not sufficient to deduce the existence of life. A motor car makes ordered progress along a road out of the disordered motion of the exploding fuel in its combustion chambers, but it is certainly not a living thing. Yet it has attributes which we associate with living, as is acknowledged in the common expression for an engine 'coming to life' or being described as 'dead' when it will not start. But even if local pockets of extremely high order are not sufficient for life, they are certainly necessary. The sorting process—the creation of order out of chaos—against the natural flow of physical events is something which is essential to life and has the advantage that it can be subjected to quantitative analysis. The main general problem of finding a physical basis for the material aspects of life is to discover the mechanism by which such a high degree of sorting is produced within the framework of the established laws described above, in the very special physical conditions which pertain in the biosphere. Natural selection provides a partial answer. Consider a system which reproduces itself during a finite lifetime. Any minor random changes in structure which tend to prolong the life of the system (or in some other way increase the average number of offspring) will tend, in the course of time, to predominate in the species. In this way purely random fluctuations producing a type of order which is favourable to the race, get trapped and are handed down to subsequent generations. But it is hard to see how this operates in the very early stages of development. It is also hard to see why it has led to the evolution of life forms of ever-increasing complexity. If survival is the essential characteristic for trapping fluctuations, very simple organisms would appear to be just as well, if not better, equipped than complicated ones.

Another aspect of this question is that one of the main functions of the brain is the ability to recognise and to create order. We have seen how the passage of time is indicated by the mounting disorder in a living-room occupied by a small, unthinking child. If, after seeing a room in chaos, it is subsequently found in good order, the sensible inference is not that time is running backwards, but that some intelligent person has been in to tidy it up. If you find letters of the alphabet ordered on a piece of paper to form a beautiful sonnet, you do not deduce that teams of monkeys have been kept for millions of years strumming on typewriters, but rather that Shakespeare has passed this way. The entire industrial revolution is based on the ability of the human brain to engineer local pockets of ordered motion.

It is not inconceivable that the physical processes, which underlie the mental activities of deduction and decision-making may also be connected with ordering. The decision which you, the reader, must now make whether to stop reading this book at once or continue to the end of the chapter may be mirrored in the physics of your brain by some such ordering process.

These ideas take us round in a very interesting circle. The ability of the mind to sort order from chaos shows itself in its most sophisticated form in mathematical reasoning. We have picked out the sorting process as being typical of animate rather than inanimate matter. Why is it then that when we come to examine the inanimate world we find it controlled by laws which can only be put in mathematical terms?

For classical physics this may not be so surprising since the axioms of arithmetic and geometry are based on the physical processes of counting objects and measuring distances. Even differential calculus is a direct attempt to put physical notions of velocity and acceleration into precise terms. But it is much harder to understand how it was that when a mathematical description of physical measurement was needed the mathematical theory of operators, already developed in the abstract by mathematicians, proved to be exactly what was required. What does an atom know of eigenvalues or a hadron of the theory of unitary groups? It was this question which led Sir James Jeans to make his much quoted statement that 'the Great Architect of the Universe now begins to appear as a pure mathematician'. This remark is not as profound as it looks. It was a flamboyant and picturesque way for Jeans to express his justifiable surprise that quantum theory had worked.

Whatever the answers to these questions, it is surely true that one of the greatest achievements of the human mind through all time, in terms of the discovery of order in circumstances of apparent chaos, is the finding of the physical laws which underlie the complex wonder of the Universe and determine in qualitative terms its ultimate destiny. The frontier of this continuing endeavour now lies in the rich and exciting world we have discovered at super-high energies at the heart of the atomic nucleus.

Index

INDEX

INDEX

INDEX